Homeowner's Guide to Appliance Repair

Karl Michael Baymor

Ideals Publishing Corp.
Milwaukee, Wisconsin

Table of Contents

ISBN 0-8249-6125-0

Published by Ideals Publishing Corporation
11315 Watertown Plank Road
Milwaukee, Wisconsin 53226

Editor, David Schansberg

Cover photo by Jerry Koser. Materials courtesy of Elm Grove Ace Hardware.

Introduction to Home Appliance Repair

Enthusiastic do-it-yourselfers and home repair hobbyists have long repaired their own appliances. In the past, many of these do-it-yourselfers enjoyed tinkering with electrical appliances for the enjoyment of learning how they work. Of course, money savings have always been a motivation for getting involved in home appliance repair. Today, however, saving money is no longer just an added sidelight to home appliance repair. With the cost of professional appliance repair increasing in recent years, the homeowner can save a great deal of money by repairing his/her own appliances. How many of you have ever had an appliance repaired, only to find out that the cost of the repair was almost as much as or even more than what a new replacement appliance would have cost? Check any of your old appliance repair bills and you will probably find that a big part of that bill went toward the cost of labor, labor that would have cost nothing if you had done the job yourself. With today's rising appliance repair costs, anyone who wants to cut down on basic home maintenance costs should seriously consider doing his/her own appliance servicing. Not only will you be able to save money, but you will often be able to get your appliance repaired more quickly than would be the case if you left it with a service representative. Also, you will have the satisfaction of knowing that you were able to complete an intricate task—a reward of its own. The truth of the matter is that home appliance repair is not that difficult to do if you have a little bit of knowledge about how appliances work and some basic information about what to look for when something goes wrong.

This book is not designed to make you the equivalent of an expert professional appliance repairman. But, if you use the book, go out and buy a few necessary tools, and make a minimal effort, you are going to be able to avoid many costly repair jobs. Sometimes the simplest problems can cause an appliance to stop working, but you will still have to pay a professional repairman a lot of money to get it back in working order if you do not try to do the job yourself.

To help you to get started, this book first goes over the tools you will need for appliance repairs. This is followed by an introduction to the basics of electricity and appliance motors, plus some tips about general repair techniques. Then the major portion of the book is devoted to the common problems of more than 15 separate small appliances and seven large appliances. Illustrations and easy-to-read troubleshooting guides are used to help make diagnosing and repairing malfunctioning appliances as easy for you as is possible. You can also help yourself by reading and using the service manuals that are published by appliance manufacturers and usually come with an appliance when it is purchased. Too often consumers throw these manuals out with the appliance packaging. Get in the habit of reading over them when you buy an appliance and keep them all together just in case you have problems later.

Tools and Test Equipment

The homeowner who is going to repair appliances will need the proper tools to perform the work correctly and efficiently. Some of the common tools you will need are described below. Some should be considered primary, while others are more optional. The primary, standard, multipurpose tools that are especially valuable for home appliance repair are as follows:

- Drill: ¼-inch or ⅜-inch portable electric.
- Extension cord.
- Extension light.
- Flashlight.
- Knife: pocket or electrician's.
- Nut-driver set.
- Pliers: channel lock, combination slip joint, diagonal cutting, and needle-nose.
- Screwdrivers: conventional (set), offset, Phillips-head (set), and screw-holding.
- Wrenches: adjustable or Crescent, open-end and box or combination (set), setscrew or Allen (set).

These tools are fairly common in the home workshop and their functions are self-explanatory, so we will not go into detail here; however, the tools in the following list are a bit more specialized and will be briefly described:

- Crimper/wire stripper.
- Lamp voltage tester/test light.
- Jumper wires.
- Soldering iron.
- Volt-ohm-milliammeter (VOM).

Electric Terminal Crimper/Wire Stripper Also called the multipurpose tool, the crimper is one of the best all-around tools to have for repairing appliances. It is used to measure wire sizes, effectively strip insulation from wires, cut and splice wire, and, as its name implies, to crimp insulated solderless terminals to ends of wire. Crimpers are also equipped with special jaws for cutting small bolts or screws. This tool is absolutely essential for repairing many appliances.

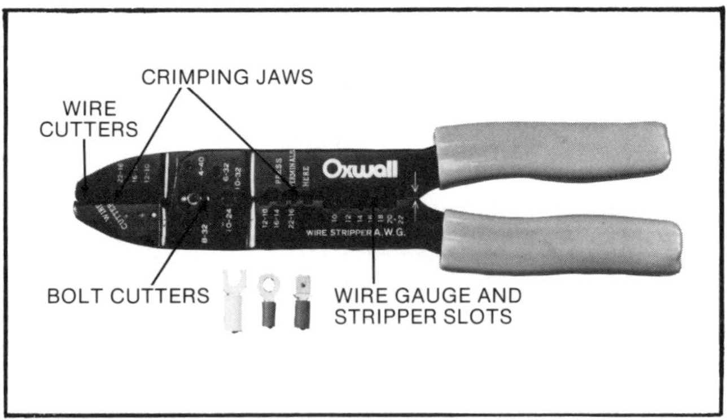

The electric terminal crimper is one of the best all-around tools to have for appliance repair.

Wire is cut by placing it between the cutting edges and simply squeezing the crimper handles. The amount of insulation to be stripped off the end of a wire can be measured by using the wire gauge on the side of the crimper. Then the wire is put into the proper size stripping slot. Close the crimper handles; the stripper slots cut through the insulation but will not cut the wire. With the handles closed, pull the wire so that the insulation is stripped off. Crimper jaws are located at the top of the crimper tool. Place the desired solderless terminal into the correctly sized grooved jaws for the gauge of wire being used. Then crimp the terminal to the wire. A bolt or screw can be cut by threading it into the special cutting jaws of the tool. The end of the bolt or screw exposed, plus the thickness of one jaw of the tool—usually ⅛-inch—will be the length of the portion cut off. To cut, squeeze the handles together.

Lamp Voltage Tester This is also sometimes simply called a test light. It is used to quickly test for voltage, although it is not by any means a substitute for the all-purpose volt-ohm-milliammeter. It is often necessary in appliance repair to find out where a problem exists in an electrical circuit. For example, in a typical appliance, the current comes in at one side of the plug, goes through the wiring, switch, motor, and/or heating element, and then back to the other side of the plug. If there is a break in the circuit—turning the switch off is one way of doing this—the appliance will not operate. A test light is used to determine if an electrical circuit is complete. There are four common variations of this simple but useful tool: an ordinary incandescent light bulb, an incandescent light bulb on a live system, a neon lamp, or a flashlight bulb.

The ordinary light bulb type of a test light is the most common. A 25-watt, 120-volt bulb may be used as a test lamp to check 120-volt current, while a 25-watt, 240-volt bulb should be employed when checking appliances which operate on a higher working voltage. A 120-volt bulb burns out very quickly on a 240-volt circuit. But, it is rather difficult in some parts of the country to find a 240-volt bulb; they must be purchased, as a rule, from electrical supply houses as a special order. Use a keyless weatherproof pigtail socket of molded rubber construction with leads permanently attached. To each of these permanent leads, a test probe must be attached. These probes can be purchased or made from wire.

An appliance is plugged in and these probes can be touched across any two wires to determine if there is voltage between them. If there is voltage, the light bulb will glow. If the bulb does not light up, there is little if any voltage. The light test is primarily used to look for loose connections, switches that will not carry current, open thermostats, broken

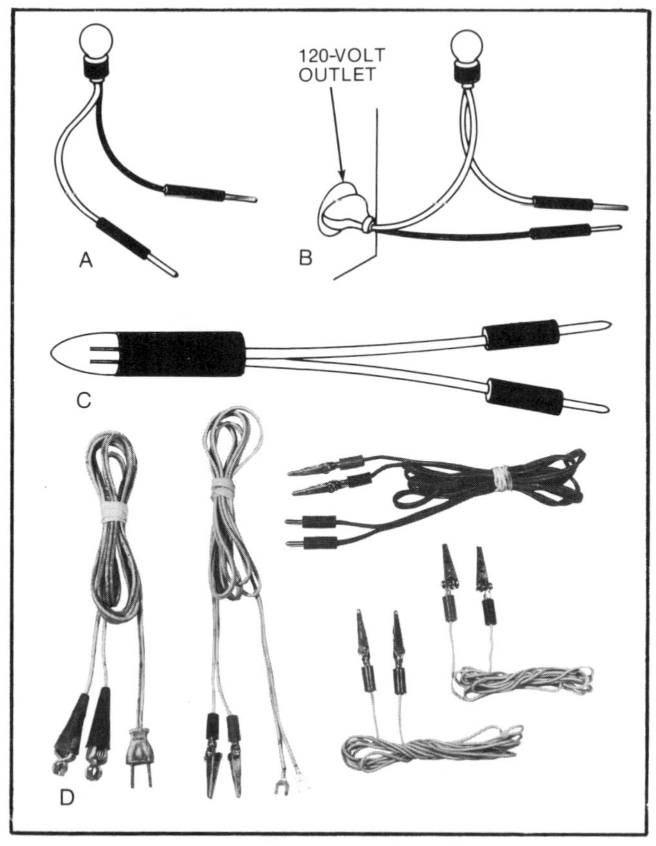

Three types of test lights used to make quick voltage checks on circuits: (A) incandescent test light, (B) live test light, and (C) neon test light. (D) Jumper wires used to bypass suspected malfunctioning parts in appliance electrical circuits.

wires, and burned-out lamps. For finer testing, you will need more sophisticated meters.

Jumper Wires Jumper wires or test cords are used to bypass suspected malfunctioning parts of an appliance and to check power cords. There are many variations of these inexpensive—you can make your own—and frequently used tools. As you begin to repair more appliances, you will be able to determine what types and sizes of jumper wires you will need.

Before connecting jumper wires, always be sure that the appliance power cord is disconnected from the power source, whether convenience outlet or battery. Connect the jumper wires, making sure that the clips or probes are not causing short circuits. Then reconnect the power cord plug. If, for example, you suspect that the power cord of an appliance is damaged, causing an open circuit, first disconnect the appliance power cord from the power source, then connect the jumper wires as a bypass in place of the actual cord. This test bypass is then connected to the power source. If the appliance works, you know that you have found the problem: the appliance power cord is defective. An advantage of using jumper wires is that they can be made to any length, allowing large areas of a circuit to be checked in one shot.

Soldering Iron A soldering iron is used in appliance repairs to solder electrical connections between wires, terminals, components, circuit boards, and connectors. The miniature or pencil iron and the instant-heat soldering gun are the best for home appliance repair. Soldering irons and solder made of lead and tin are used in low-temperature and motor-type appliances, while brazing equipment and a silver alloy solder that has a higher melting temperature are used to join electrical connections in larger appliances—refrigerators, for example—and higher-heat producing appliances—clothing irons, for example.

Basically, the process of soldering, called brazing in the case of silver soldering, consists of physically joining two connections to make a strong mechanical joint. First, all parts to be soldered must be cleaned or the solder joint will not hold. The soldering iron or brazing equipment is used to heat this joint until the temperature reaches the melting point of the solder: 360 degrees to 600 degrees F for a lead-tin solder and 1,100 degrees to 1,200 degrees F for silver solder. First, the joint is heated with the soldering iron, then a solder flux is applied which will clean the joint and allow the heated solder to flow freely. Next, the solder is touched to the joint as the iron heats it. The solder should be touched to the joint on the side opposite the iron, causing the solder to melt and to flow toward the heat of the soldering iron, filling the joint. When

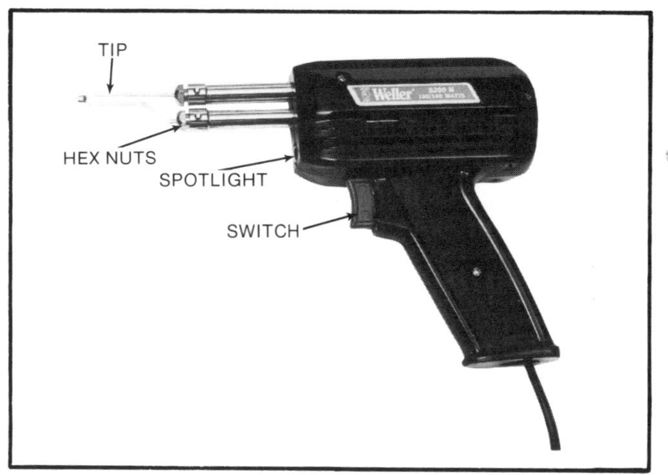

The instant-heat soldering gun is used for heavier-duty soldering.

the heat is removed, the solder cools and forms an electrically conductive joint. Take note that rosin-core solder is always used for electrical connections. Never use acid-core solder when repairing electrical parts.

Brazing equipment consists of a small tank of acetylene, a pressure regulator and torch, solder, flux, and a flint spark lighter. The acetylene is mixed with the oxygen in the air to produce a hot, clean flame. The pressure regulator reduces the tank pressure of up to 250 pounds per square inch to a regulated pressure of 10 pounds per square inch at the torch tip. The flint spark is a simple and safe way to start the flame. The silver solder or brazing alloy should be 45 percent silver; also, a cadmium-free alloy should be used because cadmium produces harmful fumes. The flux used must be suitable for brazing copper to copper, copper to brass, and copper to steel. This information will be on the flux package.

Volt-ohm-milliammeter A volt-ohm-milliammeter (VOM), which is sometimes called a multimeter or multitester because of its versatility, is an essential tool for anyone repairing appliances. It is also the most sophisticated tool you will need in your home appliance repair workshop. The VOM is used to test internal circuits, as well as components in an appliance. Depending upon the model used, a VOM can be used to measure AC and DC currents, AC and DC voltages, and resistance. Decibels—a measurement of audio power—as well as temperature can be read on some of these meters.

The VOM is a portable instrument that uses internal batteries where necessary for proper meter functioning. Two probes connect to the meter and to the circuit for making measurements. A rotary selector switch selects the voltage, current, or resistance function and range desired.

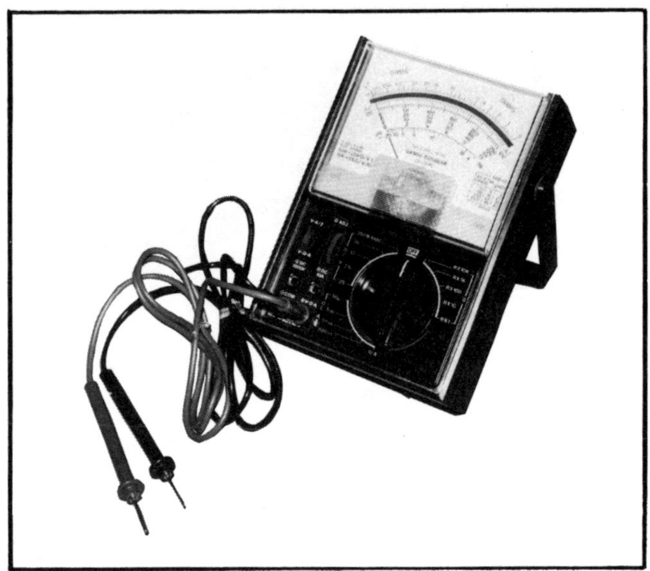

The volt-ohm-milliammeter (VOM) can be used to measure resistance, AC and DC voltage, as well as AC and DC current in some models.

The specific uses of a VOM are many. For example, if a power cord is being checked, the VOM would be set to the R × 1 resistance scale. Jumper wires are connected to the appliance end of the cord and the VOM probes are attached to the cord prongs. If the power cord is in working order, the VOM needle should move to the right, signifying zero ohms, the unit for measuring resistance. If it does not, the cord is faulty and must be replaced. The advantage of using a VOM for a continuity test rather than a continuity tester is that the VOM not only determines whether the circuit is open, but also whether the proper amount of resistance exists in a component. Remember to avoid leaving the meter function/range selector switch at the resistance function for long periods of time, because the current from the battery will be drained at that position.

We have already mentioned that the VOM can also usually measure AC and DC current and AC and DC voltage. For example, to check voltage in a wall outlet, a properly rated VOM is dialed to a position indicating 120 to 240 AC volts and its probes are inserted into the outlet. The meter will indicate the presence of voltage and the amount of voltage if the outlet is hot. Often, it is necessary to test for voltage drop in a wall outlet to help determine why an appliance is not working properly. The VOM can easily determine if any potential problems exist. By turning the indicator to a DC voltage reading, the VOM can be used to trace voltage in circuits and to work as a battery tester. Although the VOM combines several functions, it can be supplemented by an ammeter to measure current if your VOM does not have this capability, and a wattmeter for measuring electric power in watts.

Miscellaneous Tools Besides the tools already listed, other tools commonly found in the home workshop are often useful for appliance repair. Some of these optional items that you might consider obtaining if you do not already have them are as follows:

- Awl or scriber.
- Chisel, flat.
- Clamp, spring.
- Files, metal.
- Hacksaw (for cutting sheet metal, tubing, screws and bolts, steel bars, and other metal).
- Hammers, ball peen and soft-faced.
- Magnet (for retrieving parts that have dropped into an appliance).
- Penetrating oil (for releasing frozen machine screws and nuts).
- Punches: starter, pin, and center.
- Riveter (hand) and rivets.
- Socket wrench set with ratchet.
- Tap set.
- Toothbrush (as a cleaning tool).
- Vise, machinist's.
- Vise grips.

None of these tools is usually absolutely essential to repair most appliances; however, many of them have uses particularly suited to appliance repair, which can make your work much quicker and easier.

Basics of Electricity and Appliance Motors

An electric current is the flow of electrons along a conductor caused by the application of an electromotive force. This flow is measured in amperes, the fundamental unit of current flow or in other words the rate of flow of electricity. The ampere and the flow of current can be compared to the amount of flow of water past a specific point in a given time, that is, the rate of flow of water. Electrons, however, do not actually flow along a conductor from one end to the other. Rather, each free electron only travels a short distance and then bumps the next electron, which in turn also travels a short distance and bumps into another electron.

You will encounter two types of current flow in appliance repair work: direct current (DC) and alternating current (AC). Direct current is the flow of current in one direction only. Examples of direct-current flow are a flashlight or an automobile battery. Electrons leave the negative terminal on the battery, flow through the circuit, and return to the battery at the positive terminal. This flow of electrons is always in the same direction. Alternating current, however, reverses its direction at regular intervals. First, it peaks in one direction, reverses to zero, then peaks in the other direction. Picture a water pump that drives the water clockwise, then very quickly reverses its pumping direction and moves the water counterclockwise. In electricity, the direction changes very rapidly. For example, the alternating current of home electricity changes direction 60 times (60 hertz, Hz) per second. Most appliances are powered by alternating current provided by convenience outlets in the home. Truly portable appliances are powered by batteries which are a source of direct current; most electric toothbrushes, for example, have rechargeable batteries. But while most appliances are powered by alternating current from the house current, this alternating current is often rectified or converted to direct current within the appliance.

We have already stated that the ampere is a unit for measuring current flow. The ohm (Ω) is a measurement of the resistance offered to that flow of electricity. Resistance is the property of a substance that resists the flow of electrons. It can be compared to the flow of water in which the factors that influence the flow are the inside diameter of the pipe and the smoothness of its inside surface. Every material offers at least some resistance, whether to water or electricity, including electrical conductors such as copper and aluminum. This resistance to flow, or friction, increases as velocity increases. You will find that the wire used to carry a heavier current has a thicker diameter to reduce the resistance of the wire to the flow of electrons. In other words, a short, large-diameter wire will carry more electrons than a long, thin one. The thicker and shorter a wire is, the better is its ability to carry heavy currents. An ohm of resistance is equal to the amount of opposition offered by a circuit or conductor to the flow of one ampere of current when a pressure of one volt is applied across its terminals. The resistance value of a conductor will primarily depend upon four factors: the nature of the material, the size of the conductor, the length of the conductor, and the temperature of the conductor. Metals, especially copper and aluminum, are good conductors, as are large diameter, longer, and cooler conductors.

Voltage is the electromotive force that will produce a current of one ampere through a resistance of one ohm. In other words, a volt (V) is equivalent to the electric pressure required to force one ampere of current through a resistance of one ohm. Again, using an analogy with a water system, this electromotive force or pressure resembles the pressure caused by a water pump; in the case of an electrical system, the flow is electrical current.

Power is the rate at which electric energy is generated or used. Power requires voltage and current, that is, electric pressure accompanied by a flow of electrons. Without voltage there can be no current. Without current, no work can be done or energy used. Power is the product of voltage and current (volts × amperes). A watt (W) is the unit of power. It is the measure of the rate at which electric energy is generated or expended. An applied voltage of one volt and a current flow of one ampere results in one watt of power.

An electric circuit is a path or a group of interconnected paths capable of carrying electric currents. The three basic types of circuits are the series circuit, the parallel circuit, and the series-parallel circuit. The series circuit has only one path for the flow of current. The parallel circuit has two or more paths for current flow. The series-parallel circuit is a combination of series circuits and parallel circuits. A circuit starts at the source of electricity and it ends at the source of electricity. In this circuit, or path, there are usually one or more electrically operated devices, or electric components.

Basic Electrical Components Before taking a look at appliance motors, some general electrical components should be studied. The components that will be dealt with fall into two groups: those that alter the flow of electrons in a current in some way—resistors, capacitors, inductors or coils, transistors, diodes, rectifiers, and batteries; and those that do not have much effect on electron flow—plugs, sockets, line cords, fuses, circuit breakers, and switches.

Resistors A resistor is a component that is especially designed to impede or resist the flow of electrons through an electrical circuit. Resistors are used to aid in the operation, protection, and current control of an electrical circuit. The physical size of a resistor usually determines its wattage; the larger the resistor body, the higher the current it can carry. The ohmic value of resistors may be fixed or variable. Fixed resistors range in value from a fraction of an ohm to more than 10 million ohms and cannot be changed. Variable resistors, sometimes called potentiometers or rheostats, usually cover a specified range of ohmic values. Resistors are made of wire, carbon composition, or film. The resistance value, tolerance, and wattage ratings of a wire-type resistor are usually printed right on the resistor. Carbon-composition and film-type resistors have coded color bands on them to readily identify their tolerances. When it is necessary for you to replace a resistor in an appliance, you should replace it with a resistor having the same ohmic and wattage values. However, if you do not have the exact replacement, you can obtain the value by adding two or more resistors in series or in parallel.

Plastics, rubber, ceramics, mica, glass, and wood are examples of materials that have virtually infinite resistance. They are nonconductors of electricity and do not support current flow. This sort of material is known as an insulator or dielectric. Parts such as wire covers and standoffs—used to hold coils of bare heating element or wire away from metal cases—are typical examples of insulators used in small appliances.

Capacitors and Inductors A capacitor or condenser is a device capable of storing electrical energy. When the capacitor is charged, potential energy is stored in it as electrostatic energy. This electrostatic or potential energy is capable of doing work when the charge is released through an external circuit. The principle of a capacitor involves the movement of electrons within two conductors which are insulated from each other and incorporates an electric charge induced without contact. A simple capacitor, for example, would require two conductors, plates of metal or metal foil, which are separated from each other by an insulator such as air, glass, wood, rubber, or paper. When a capacitor needs to be replaced in a circuit, replace it with the same capacitance value and with the same or a higher voltage rating.

Inductors are coils of wire, sometimes wrapped around a mold or form. When a current flows through the coil, the electrons cause a magnetic field to build up in proportion to the amount of current flow. Therefore, in a way, an inductor stores energy in the form of a magnetic field rather than an electric field, the way a capacitor does. When a material such as some form of iron or steel is placed in the center of the core, the inductance increases. A coil has an inductance of 1 henry (H) when a current, changing at the rate of one ampere per second, induces an average of one volt.

Transistors and Diodes A transistor is a device used in electrical circuits to amplify a signal or to act as a variable switch. A transistor has three terminals, called the emitter, base, and collector. Electrons flow from the emitter, through the transistor, and out of the collector terminal. The base provides a means of controlling the amount of electron flow through the transistor from emitter to collector. Therefore, the base can turn on or off the flow of electrons through the transistor. If it is necessary to replace a transistor in an appliance control system, be sure that the emitter, base, and collector are connected to their respective connections. If you install a transistor improperly, you will probably destroy it as soon as the appliance is reconnected to a current.

Diodes or rectifiers are used to rectify, or change, alternating current to direct current. The diode has the property of allowing electrons to flow easily in one direction while opposing the flow of electrons in the opposite direction; therefore, when an alternating voltage is applied to a diode, it allows electrons to flow in one direction, but opposes the flow in the opposite direction. In other words, the diode rectifies or changes the input alternating current into a direct current.

A diode has two terminals called the anode (+) and the cathode (-). Electrons flow through the diode from the cathode to the anode, negative to positive. The diodes will be marked with a band or in some other way to indicate the cathode side. It is important that replacement diodes be placed back into the circuit with the cathode ends connected to the same points as the original diodes.

Batteries A battery is essentially an energy-conversion device that converts chemical energy into electrical energy. A battery is a source of direct current. Electrons flow from the negative terminal of a battery through an external circuit to the positive terminal to power small appliances and other portable items. There are a variety of battery types. The common primary cell battery used in flashlights, some electric toothbrushes, and so on cannot be recharged or rejuvenated. When they are dead, they must be replaced. Batteries that can be recharged in some way are called secondary cell batteries. Perhaps the most obvious battery of this type is the common lead-acid automobile battery. However, the lead-acid secondary storage battery is not practical as a source of direct current for appliances, because it must be unsealed to vent gases during normal operation. Because it is unsealed, the liquid electrolyte—diluted sulfuric acid—in the battery can spill out.

Nevertheless, there are two types of rechargeable batteries that can be used in appliances: nickel-cadmium batteries and alkaline secondary batteries. Both of these battery types are hermetically sealed and therefore spill-proof. The nickel-cadmium battery can be repeatedly recharged, is maintenance-free, and can be mounted or used in any position. Its average operating voltage is 1.2 volts. Nickel-cadmium batteries are generally expensive but durable. The alkaline secondary battery cannot be recharged as often as the nickel-cadmium battery, but it costs a lot less. Also, these batteries must not be allowed to discharge completely or they will be dead, and you will not be able to recharge them any more.

Other Electrical Components All of the components we have just briefly discussed affect the flow of electrons in an electrical circuit. Another group of basic electrical components is common to many appliances. These common components are plugs, sockets, line cords, fuses, circuit breakers, and switches. Because these components are factors when repairing many different appliances, we will review them here before discussing specific appliances.

Plugs, Line Cords, and Sockets Plugs, line or power cords, and sockets are the basic instruments for moving electrical current from the house current to the appliance. Of course, a typical power cord—an extension cord, for example—has a plug at one end and a socket at the other. However, many appliances do not have a detachable socket. Instead, with an outlet plug on one end, the line cord passes through the appliance cover where it is held in place by a plastic strain reliever, by a knot which is unable to pass through a grommet in the cover, or by a clamping device.

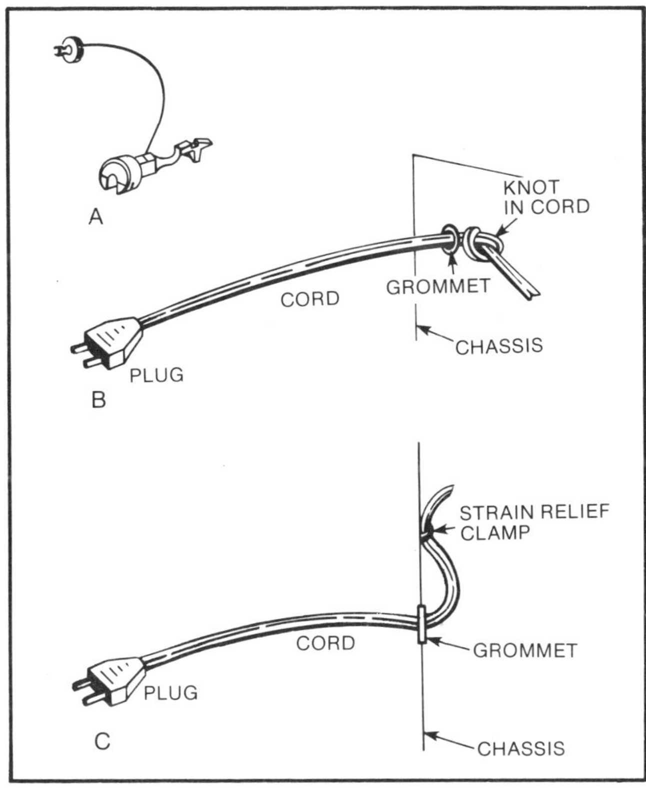

Three methods of holding an appliance line cord in place: (A) a strain reliever, (B) a grommet and a knotted cord, and (C) a clamping device.

If an appliance is not working properly and a continuity test determines that the problem is in the line cord, you must replace the cord. Furthermore, if you ever notice that an appliance cord has developed frays, nicks, or small breaks, you should replace the cord for safety's sake, regardless of whether the appliance works. If the cord is not detachable, this will necessitate opening the appliance case, unsoldering the wires, and replacing the old cord with a new cord and plug. Be sure to replace the strain reliever or the clamp, or tie an Underwriter's knot within the appliance.

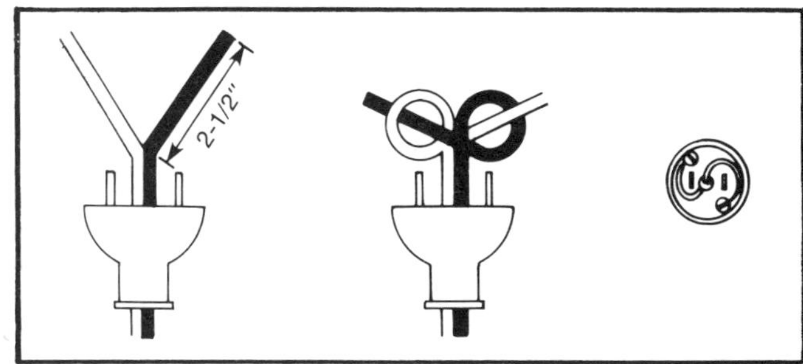

The Underwriter's cord knot provides strain relief for wires in a plug or socket connection.

If the problem seems to be in the plug, check for loose connections, broken contacts, stripped screw threads, and a broken housing. If the line cord wires attach to the plug at screw posts, check the posts. Make sure that the posts are bright and shiny—if not, clean with steel wool—and are screwed tightly against the wires. Also make sure that there are no loose strands of one wire in the vicinity of the other connection. Loose strands of wire should be corrected by unloosening the post and twisting the wire strands together. You can ensure that no strands will become loose by applying a small amount of solder to the end of the wire. Insert the wire through the plug and tie an Underwriter's strain relief knot. Then wrap the wire around the post in a clockwise direction and tighten the post; the black wire (hot wire) of the appliance is connected to the smaller plug-prong contact if the plug is so designed. Visually inspect the rewired plug to ensure that no short circuit exists between the wires (the wires do not touch each other) and that the wires are not frayed.

When you replace a line cord, replace it with the same size and type of cord. There are a number of cord types, but three types are most prevalent: zip-cord, SV cord, and heater cord. All line cords have stranded wire inside them; 7 to 10 strands is common, but the number of strands may be as high as 65. Stranded wire is usually used on appliances because of its flexibility.

If there is a socket on the line cord, check it for loose connections, broken or damaged contacts, stripped screw threads, and a broken housing. If the cord into the socket is replaced, secure the cord with the socket's built-in clamping device; in the absence of a clamping device, tie an Underwriter's knot. Strands of wire should be twisted firmly together; a little solder will hold them firmly for a good connection. Wrap the wires around the screw-posts in a clockwise direction and tighten the screws securely. Reassemble the socket halves.

You can test a line cord assembly, the cord itself, a plug, or a socket by means of continuity testing. In making continuity tests, ensure that the plug is removed from the convenience outlet before making

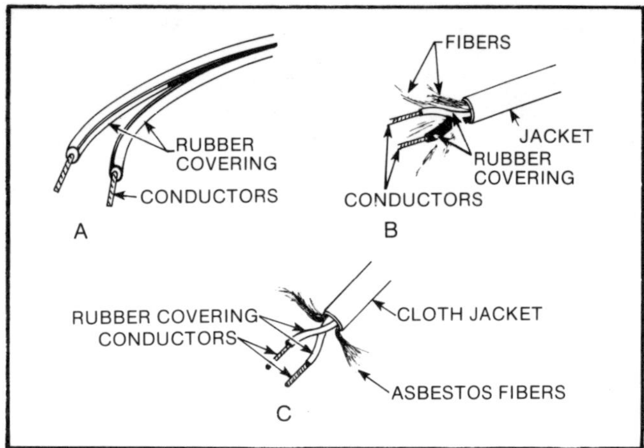

Three types of line cord commonly used on appliances: (A) Zipcord is for low current appliances. (B) SV cord is used for moderately heavy currents. (C) Heater cord is used where an appliance draws more than 5 amperes of current.

a test. Adjust the dial of a volt-ohm-milliammeter for resistance testing and connect one probe to one end of the part being tested and the other probe to the other end of the part. An indication of zero ohms indicates that the part is defective and must be replaced. When continuity-checking a line cord, bend the line cord back and forth, especially near the plug or socket, to check for internal wiring breaks that may only be open when the cord is bent. Intermittent readings other than zero ohms indicate a defective line cord.

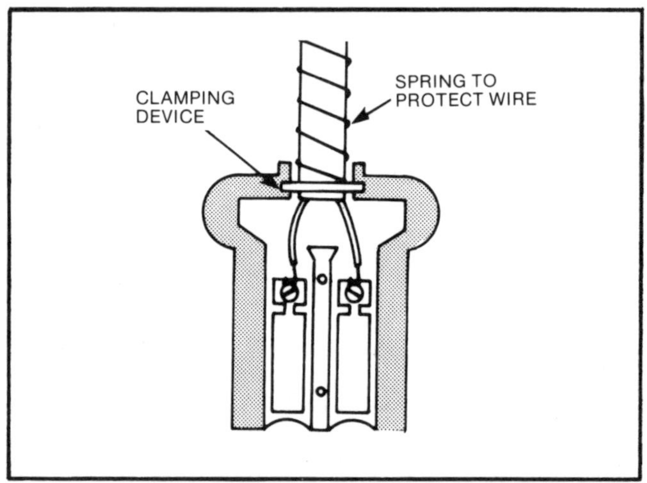

Sockets, as well as plugs, should be checked for loose connections, broken or damaged contacts, stripped screw threads, and a broken housing, if an appliance does not work and the line cord is suspect.

In electrical wiring, wires are identified by sizes; however, the smaller the wire size number, the larger the wire. In addition to the wire size number, wires are also rated according to the load they can handle—both the normal load and the capacity load.

Normal and Capacity Loads for Wiring Sizes Commonly Found on Appliances

Wire Size	Normal Load	Capacity Load
No. 10	27.5 amps (2,400 W)	30 amps (3,000 W)
No. 12	16.6 amps (1,900 W)	20 amps (2,400 W)
No. 14	12.5 amps (1,500 W)	15 amps (1,800 W)
No. 16	8.3 amps (1,000 W)	10 amps (1,200 W)
No. 18	5.0 amps (600 W)	7 amps (840 W)

The nearer a wire gets to its capacity load, the greater the chance of some problem developing. When a wire is near its capacity, an increasing amount of heat is generated which could cause a fire. Most small appliances operate on either No. 16 or No. 18 wire. Larger appliances usually require No. 10, 12, 14, or 16 wire.

Extension Cords Extension cords come in various lengths, which is actually an important factor when selecting one. The longer the cord, the greater the distance an appliance can be from a convenience outlet; however, as we mentioned earlier, the longer a cord is, the greater the chance of a voltage drop. Although there is a certain voltage range within which appliances will operate without causing damage to a motor, the appliance will probably not operate correctly if the voltage drops below the allowable tolerance. To ensure that the length of an extension cord will not drop the voltage below the range of tolerance, check with the extension cord table here.

Extension Cord Tolerance Ranges by Wire Size

	Load to 7 amps	From 7 to 10 amps	10 to 15 amps
To 25 ft.	No. 18	No. 16	No. 14
To 50 ft.	No. 16	No. 14	No. 12
To 100 ft.	No. 14	No. 12	No. 10

Fuses and Circuit Breakers A fuse is a protection device. It consists of a piece of wire or a strip of fusible metal that melts when the current becomes excessive. This fuse is connected in series with the hot line of the appliance. If a short circuit or other malfunction occurs within the appliance, the current increases and as it passes through the fuse, it causes the fuse to open up, preventing any further flow of current. There are various types of fuses, although a fuse can consist simply of a single strand of fine wire.

The proper functioning of a fuse can be checked by two methods. The first method is to perform a continuity test on the fuse. Remove power from the appliance and then remove the fuse from the circuit, dial your VOM to a resistance function, and touch the probes to the two contact parts of the fuse; this could be the two ends or the center and the screw-in threads. If the meter indicates zero ohms, the fuse has continuity and current passes through it. If the meter indicates infinity, the fuse is open or burned out and must be replaced. It is also a good idea to check and see why the fuse was blown. If a malfunctioning appliance caused the fuse to burn out and the problem is not corrected, a replacement fuse will probably also burn out.

A circuit breaker is also a protection device, similar to a fuse, except that the circuit breaker is popped or moved to the OFF position, rather than being destroyed by excessive current. The circuit breaker must be reset to the ON position and the appliance will operate again. This is assuming that the appliance did not malfunction or that a malfunction that caused the excessive current has been repaired. In actual practice, circuit breakers and fuses are seldom a part of the design of an appliance; rather, circuit breakers or fuses are in the power-distribution system of a home, apartment, or industrial building.

Switches Switches are used to turn appliances on and off. There are slide switches, push-button switches, rotary switches, and toggle switches. When a switch is open or OFF, there is no path for current flow; when the switch is closed or ON, metal contacts engage, which allows the flow of current. You can easily check a switch by making a continuity test. Determine the type of switch that you have. Ensure that all power is removed from the appliance and remains removed for continuity tests. Then connect the two leads of the VOM to the two switch contacts to be checked. With the switch in the OFF position, there should not be any continuity; the VOM should indicate infinity. When the switch is placed to the ON position, the meter should indicate zero ohms, verifying that there is continuity through the switch.

Grounding Fortunately, the ground or earth has a tremendous ability to receive, dissipate, and, under some conditions, conduct electric charges without upsetting its electrical balance. Intentional grounds are those in which a conductor is intentionally connected between a source of electricity and a cold-water pipe or a metal pipe driven into the ground. They also pertain to the connecting of a conductor between the metal frame of an electrical device and a point which leads to the ground. The ground wire keeps the rest of the electric system from being subjected to a charge of electricity greater than its capacity. Accidental grounds are those in which a bare wire conducting electricity touches the frame of a device or appliance. This permits the electricity to take an unintended path and may result in injuries or damage to the appliance or device.

For safety's sake, you should always make what is called a ground test at the completion of any repair involving an AC appliance with a metal case. The ground test ensures that no bare wiring inside the appliance case is touching the case, making it hot. To make the ground test, connect the appliance power plug into the convenience outlet. Turn the appliance on. Dial your VOM to the AC voltage function and connect one lead to the case and the other lead to the plate screw on the convenience outlet. The VOM should read zero volts. If the meter does not read zero, there is AC leakage to the case. You must disconnect the appliance from the power source, reopen the case, find the cause of the leakage, correct the problem, reassemble the case, and again perform a ground test.

Appliance Motors Many of the appliances discussed in this book are driven by a motor of some type. Simply put, a motor is a device that converts electrical energy into continuous rotary mechanical energy. The electrical input to the motor is from either an alternating-current or a direct-current source. In most household appliances, the electrical input to the motor is from 115 volts to 120 volts AC; the output of the motor is a rotary twisting or torquing motion from a shaft, which is then coupled directly or through belts, gears, or pulleys to other mechanical parts in the appliance. Motors used in small appliances are fractional-horsepower motors; that is, their output is less than 1 horsepower. The output of fractional-horsepower motors ranges from $\frac{1}{20}$ to $\frac{3}{4}$ horsepower.

All motors, whether driven by electrical alternating current or by direct current, operate on the same principle of magnetism—like poles repel and unlike poles attract. This principle applies to permanent magnets used in small direct-current motors and to electromagnets in which the power is provided by an AC or a DC source. Some small appliances use both permanent magnets and electromagnets in combination, but the principle of operation remains the same.

Every motor consists of two parts that create two opposing magnetic fields; one of the magnetic fields rotates inside the other. One of the magnets that produces a magnetic field is known as the stator or stationary piece. The stator is often called the field; it can be a permanent magnet or an electromagnet. The other magnet, which generates the second magnetic field, is known as the rotor or rotating piece. The rotor is always an electromagnet; if the rotor contains windings (as in DC motors and universal motors), it is known as an armature. The stator and rotor are so constructed that the rotor revolves within the stator. The opposing magnetic fields produced by the stator and the rotor cause a rotary mechanical output. Thus, an electrical input is converted into a mechanical output.

You can probably visualize that when the rotor is in the stator and there is no movement, there is a possibility that the unlike magnetic poles will attract one another and the motor will not be able to operate. This condition could occur when initial electrical power is applied to the motor. Thus, there must be a means of getting the rotor started; this is accomplished in different ways by different types of motors, but in all of them the principle is the same. The two poles are edged apart before the motor is started. Once the poles are slightly misaligned, the rotor rotates and chases the magnetic poles of the stator by attraction and repulsion. Once running, the momentum of the rotor keeps the motor running; that is, the attraction of the unlike poles cannot stop the rotor because the momentum drives the rotor on past the point of maximum attraction and then the repulsion forces take over again.

There are a number of different kinds of motors in existence. Of the different kinds, three are used most prevalently in small appliances: series-wound universal motors, permanent-magnet DC motors, and shaded-pole induction motors. Two other types of motors, split-phase and capacitor-start motors, are infrequently used for small appliances but are common in larger appliances.

Additional Repair Techniques

We have already discussed several of the more important repair techniques you will need to know before servicing appliances: soldering and brazing, crimping, checking and replacing line cords, replacing line plugs, and continuity testing. In addition, there are several other general repairing tips that should be reviewed.

Making Wiring Connections On many appliance jobs there will be a need for splicing and terminating wires. A good electrical connection has several requirements. Wires must be free of insulating materials, connections should be secure, and splices must always be covered with plastic electrical tape, so that the wire is as well insulated as it was before insulation was removed. Take note that wire splices and joints are most often used with the solid rather than the more flexible stranded type of wire. Nevertheless, you will probably have occasion to do some wire splicing during your appliance repairs, so it is a good idea to learn the basic joints. The most common ones are the western splice, staggered splice, pigtail joint, fixture joint, and knotted tap joint.

Solderless Connectors When there will be no strain on the wire, a quick and satisfactory method of joining wires is simply to use convenient solderless connectors. Four of the most common types of solderless connectors are the split-sleeve, split-tapered-sleeve, crimp-on splicers, and the wire nut.

To connect a split-sleeve splicer to its conductor, first insert the stripped wire tip between the split-sleeve jaws. Using a tool designed for that purpose, force the slide ring toward the end of the sleeve. Close the sleeve jaws tightly on the conductor; the slide ring holds them securely. To mount a split-tapered-sleeve, strip the conductor and insert it in the split-tapered-sleeve. Turn or screw the threaded sleeve into the tapered bore of the body. As the sleeve is turned in, the split segments are squeezed tightly around the conductor by the narrowing bore. The finished splice must be covered with insulation. We have already mentioned the crimp-on splicer which is jointed by using the plier-like hand crimpers. The insulating sleeve grips the outer insulated conductor, and the metallic internal splicer grips the bare wire conductor strands.

If there will be no strain on the wire, wire nuts can be used. The wires are inserted into the plastic cap, which is screwed tight. Be certain no wire is left exposed.

Solderless Terminal Lugs Solderless terminal lugs are used more widely than solder terminal lugs. They afford adequate electric contact, plus great mechanical strength. In addition, they are easier to attach correctly because they are free of the most common problems of solder terminal lugs, such as cold solder joints and burned insulation. These lugs come in many sizes and shapes, each intended for service with wires of different size. Three of the most common types are the wedge split-tapered-sleeve, the threaded split-tapered-sleeve, and the crimp-on lugs which include ring, hook, and fork models.

Repairing Motorized Small Appliances

Small appliances are generally divided into two broad categories: motorized appliances and resistance-heating appliances. Some appliances, heater fans and a few of the personal-care items, for example, combine the two basic principles in one appliance. In a further breakdown, resistance-heating appliances can be divided into two groups: those that are controlled by thermostats or timers and those that are not. There are also three basic groupings of small motorized appliances: those with single-speed motors, those with multiple- and variable-speed motors, and battery-powered portable models. Although these subgroups are distinct from each other in terms of appliance operation, they overlap as far as individual appliances go. For example, older clothing iron models were not thermostatically controlled, while most irons made today are. Or, some fans have simple single-speed motors while others are equipped with multiple-speed motors and a variety of speed controls.

To avoid confusion we will discuss the variations of each appliance all in one section. Therefore, the section on irons will include all iron models, and the section on fans will include single-speed and multiple-speed fans. However, for your own sake, it can be useful to consider small appliances in terms of general groupings, because they will have a lot in common. For example, all battery-powered portable appliances make use of some type of rechargeable battery and a recharger. The procedures for checking, recharging, or replacing a battery pack will be very similar for all of this type of appliance. We will be looking first at the motorized small appliances.

Food Mixers

Food mixers have two or three motor-driven beaters used to mix, stir, beat, or whip ingredients in the preparation of food. With attachments, food mixer models can also grind meat, extract juice, slice potatoes, crush ice, and sharpen knives. There are two basic types of food mixer: upright with a stand and portable. Both types, however, are powered by the same means.

Food mixers usually are equipped with universal motors because of the heavy loads encountered at slow operating speeds. Speeds between approximately 300 and 1,300 rotations per minute may be controlled by a tapped-field coil, allowing three speeds, movable brush, or centrifugal-governor controls. The centrifugal governors are used more often in upright mixers than in portable mixers. A typical three-speed mixer operation is usually within the following range:

Low: 300-700 rotations per minute.
Medium: 550-1,000 rotations per minute.
High: 800-1,300 rotations per minute.

The mixer universal motor is usually mounted with the shaft horizontal. The shaft has a worm drive machined into it that drives two gears containing sockets for the attachment of the beaters. To ensure complete mixing, the rotating beaters either turn the bowl and turntable or circle inside a stationary bowl. Some mixers have a shaft on the top of the gearbox to drive accessories; some models have gearboxes that are lubricated and sealed. This lubricant should last indefinitely.

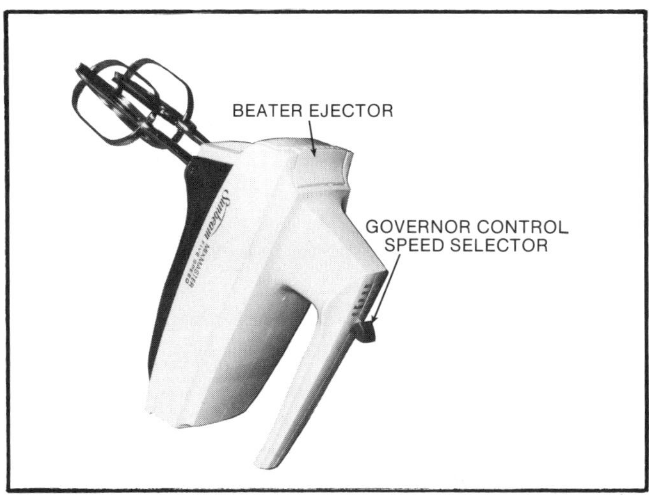

Food mixers are used to mix, stir, beat, and whip food ingredients.

When the slide switch is turned to low on a mixer having a three-speed tapped-field coil speed control, alternating current is applied through the complete tapped winding to the universal motor. Since a part of the applied voltage is dropped across each of the tapped coils, the motor runs at its lowest speed. When the slide switch is moved to medium and high, there is less voltage dropped.

When the adjustable speed control knob is rotated to the desired speed on a mixer having a centrifugal-governor speed control, it sets the spacing between the electrical contacts on the governor. Increasing

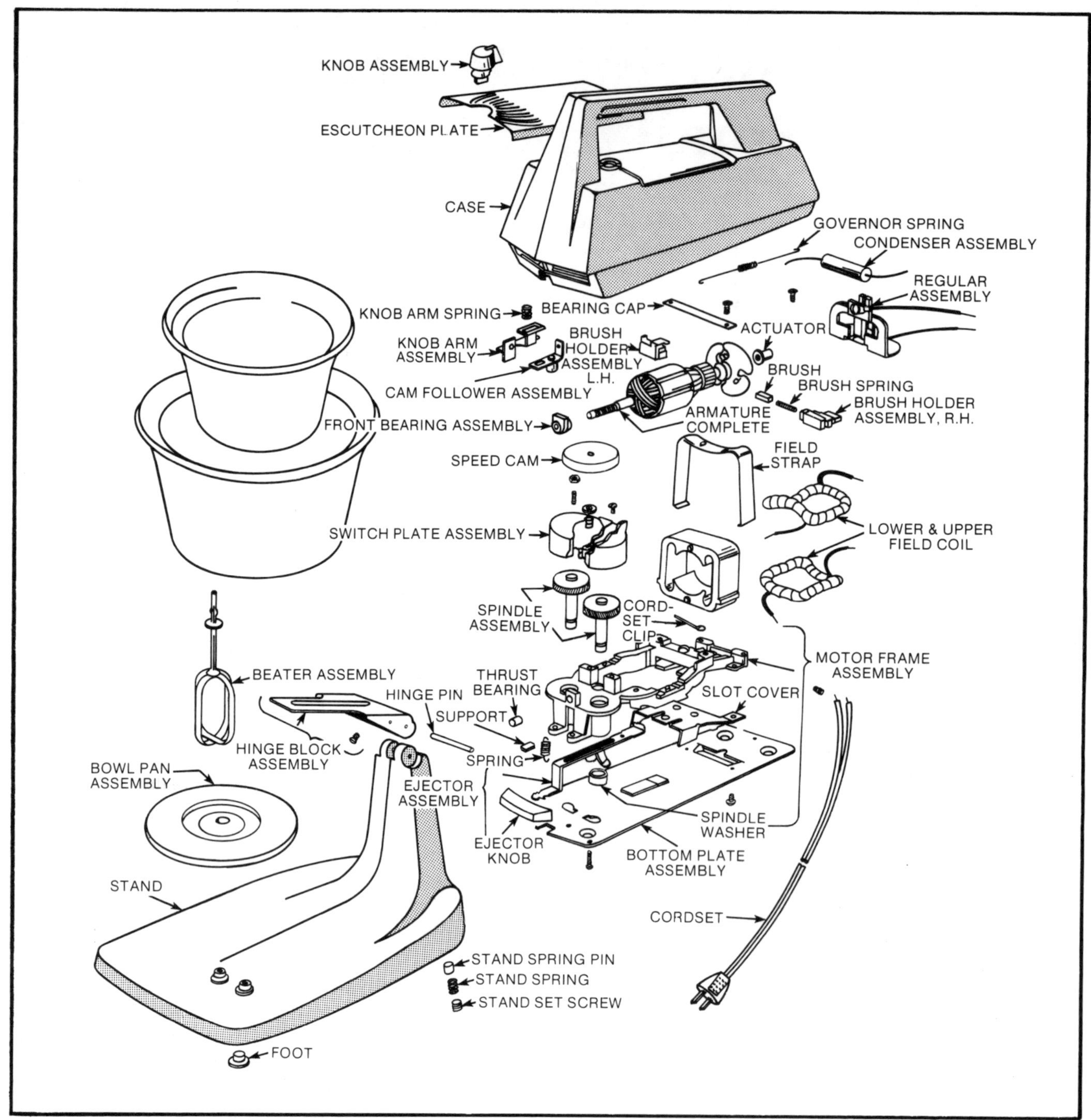

Exploded view of a typical stand-type food mixer.

the spring force holds the contacts together longer. It takes a higher motor speed with resulting higher centrifugal force to open the contacts, removing electrical power from the motor. Since the contacts are on the rotating armature, collector rings are necessary to electrically connect the contacts with the rest of the electrical circuit. The centrifugal-governor contacts are initially open with the motor not running. As soon as the control is moved, it closes the contacts and turns the motor on. Thus, the adjustment serves as an ON-OFF switch in addition to serving as a speed control.

As the motor speed increases, the governor is affected by centrifugal force, which opposes the spring tension on the contacts. When the motor speed set by the control adjustment is attained, the centrifugal force overcomes the spring tension and the governor contacts open, removing electrical power to the motor. With no power to the mixer motor, it begins to slow down. When the spring tension overcomes the centrifugal force, the governor contacts close, applying electrical power to the motor to speed it up again. This action of the opening and closing of the governor contacts occurs many

Food Mixer Troubleshooting Chart

Trouble	Probable Cause	Remedy
Motor does not run.	Line cord is defective.	Check line cord for continuity. Replace as needed.
	ON-OFF switch or ON-OFF/ speed selector switch is defective.	Check switch. Replace as needed.
	Centrifugal-governor contacts are stuck open.	Check centrifugal governor. Clean contacts if dirty.
	Motor is defective.	Check motor. Replace if necessary.
Motor does not run at speeds selected.	Selector switch is defective.	Check speed selector switch. Replace as needed.
	Tapped-field coil defective.	Check tapped-field coil. Replace as needed.
	Centrifugal governor is operating incorrectly.	Check governor; replace parts as needed.
	If the motor has a centrifugal governor and the motor runs only at maximum speed, the governor capacitor is short-circuited.	Remove one capacitor lead from the circuit. If the motor now runs at variable speeds, replace the capacitor with a new one.
Motor seems to jerk.	Resistor is open.	Check resistor with VOM. Replace if defective.
Motor brushes spark excessively.	Brushes are worn or commutator is dirty.	Check brushes; replace if worn. Clean commutator.
Motor runs, but beaters do not turn.	Gears, couplings, or worm is defective, or bearing is frozen.	Inspect, clean, and lubricate. Replace gears that are damaged; be sure to align gears properly. Find and replace frozen bearing.
Operation is excessively noisy.	Insufficient gearbox lubrication.	Lubricate.
	Gears or bearings are worn.	Check; replace as needed.
	Centrifugal switch operation is erratic.	Check centrifugal governor; replace as needed.
	Armature is unbalanced.	Check armature by operating mixer at high speed while holding it by the handle on outstretched fingertips. Unbalanced armature will vibrate noticeably.
	Fan loose on shaft.	Tighten fan. Check for bent blades; replace fan if it is unbalanced.
	Shorted or open turns in the armature.	Check; replace as needed.
Mixer operation causes interference to radio or television.	Governor capacitor is defective.	Replace capacitor.
Beaters will not eject or stay in.	Defective spindles or retaining springs.	Check for rusted or dirty spindles or retaining springs. Clean, lubricate, or replace as needed. Check for loose plate or broken or missing spring; repair or replace as needed.

Food Mixer Troubleshooting Chart (Continued)

Trouble	Probable Cause	Remedy
Bowl does not rotate on models that have a rotating bowl.	Pivot pin on bowl pan sticking.	Lubricate pivot pin.
	Beaters too low or too high.	Adjust beaters.
Mixer blows fuse or trips circuit breaker.	Short circuit.	Check line cord, switch, and motor for short circuits. Replace defective parts.
Mixer case shocks user.	Short circuit to case.	Make ground test to locate short circuit. Replace bare wires and tape terminals.

times per minute; however, you cannot actually hear or see the motor or the beaters slowing down or speeding up.

The governor capacitor prevents arcing across the governor contacts; this eliminates or at least significantly reduces interference to a radio or television and helps prevent contact pitting. A resistor allows a small part of the input voltage to travel to the motor even when the governor contacts are open; this action prevents the motor from jerking to a faster or slower speed when the governor contacts are closed and opened.

Troubles in mixers can take the form of electrical problems in the universal motor or in the control system and mechanical problems in the gears, motor, centrifugal governor, or beaters. If the motor brushes are suspected of causing trouble, they can usually be removed from a mixer by removing plastic screws on the outside of the case.

If the mixer has a centrifugal governor and causes excessive radio or television interference, the governor capacitor is probably open. If the motor operates at only full speed, the governor capacitor is probably short-circuited. If the governor contacts stick, sand them lightly with a piece of sandpaper or replace them, if necessary. The contacts may also stick in an open position because they are dirty; in this case, clean the contacts and put a drop of oil on the pivot point.

If the mixer motor runs but the beaters do not turn, the problem is in the worm, the gears, or the coupling. If the gears are frozen or encrusted with food particles, disassemble the gearbox, clean the parts with alcohol and a toothbrush, and reassemble; lubricate as recommended by the manufacturer. A drop of SAE 20 motor oil should be placed on the inner surface of each bearing. If grease leaks from the beater sockets, the bearings are worn and the grease is thinned. Replace the bearings and the grease. When reassembling the beater gears, be sure that the beaters mesh properly; they should not touch each other. The gears are often marked with timing marks to ensure proper meshing. Look for one and two dots on each gear; then set one gear with one dot opposite the other gear at the two-dot position. Noise in a mixer could be caused by insufficient gearbox lubrication, worn gears, worn bearings, or an erratic centrifugal switch operation. Bent beaters can sometimes be straightened; but if not, replace them. Dirt on the case of the mixer can be cleaned with a solution of detergent and water; however, do not get any electrical components wet.

Food Blenders and Food Processors

Food blenders and food processors are used to stir, puree, whip, grate, mix, chop, grind, blend, and liquefy fruits, vegetables, and so on. A small motor-driven rotor blade revolves rapidly, cutting, slicing, and pounding the ingredients to the desired consistency within a short time. A blender has a universal motor that is mounted below the container or bowl and operates at speeds of approximately 3,000 to 14,000 rotations per minute. The shaft of the motor extends into the container base, which contains rotor blades that perform the blending and chopping of the ingredients in the container. From 3 to 16 speed select switches are mounted on the control panel and are connected electrically to a tapped-field-winding, a solid-state, or a combination tapped-field solid-state speed control circuit. Some blenders have a combined ON-OFF switch, some have separate switches, and some have a momentary hold-on button that allows the blender to operate as long as the button is manually held in. Some blenders also have timers with up to 60 seconds of operating time. When the time is set and released, the blender operates at the speed selected for the time selected and then automatically turns off. Blenders operate at between 250 and 750 watts.

The blender container is separable from the container base to simplify removal of the contents and

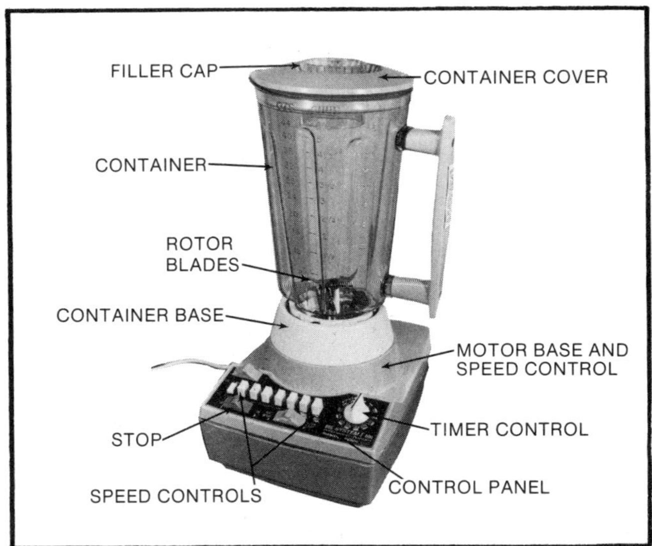

Blenders are used for mixing ingredients into a uniform composition or for grinding or chopping nuts, mushrooms, and so on.

to facilitate cleaning. The container base is separable from the motor base and speed control. The spatula, filler cap, container cover, container, rotor blades, and container base are all washable; they may be immersed and may be washed in a dishwasher. The blender motor base and speed controls are not immersible; wipe them clean with a damp cloth and keep excessive liquid off of the control panels.

Problems in blenders include electrical and mechanical malfunctions in the motor, electrical malfunctions in the speed control circuits, and mechanical troubles in the drive mechanisms and rotor blades. You can test the blender by turning it on at the slowest speed. Then press each successive speed button or rotate the speed selector; the speed of the rotor should increase with each higher speed selected. Listen for speed increases and for noises that indicate mechanical troubles. However, in multiple-speed appliances having 12 or more speeds, the speed increment between adjacent selection positions is about 300 rotations per minute. This increase in speed may not be loud enough to hear. Therefore, start at the lowest speed and select every other speed; then select the second lowest speed followed by every other speed. By checking every other speed you should be able to hear the difference.

If one speed control setting will not work or if the motor runs at a lower speed than it is supposed to, the trouble is probably a loose connection associated with the switch, tapped winding, or solid-state input. If the motor operates only starting with a certain switch along the line, the trouble is probably an open winding. If the blender has a timer, check its operation. First check for a 60-second timed operation and verify that the blender stops automatically after 60 seconds. Then set the timer to manual and verify that the blender still operates manually. Noise in a blender can be caused by a bent rotor blade or shaft; motor bearings or a blade shaft that requires cleaning and lubrication; a bent armature; or loose nuts, screws, and other hardware. Isolate the problem to either the motor or motor-driven parts. Then rotate the parts by hand to isolate the malfunctioning part. Lubricate in accordance with the manufacturer's directions; SAE 30 motor oil is usually used.

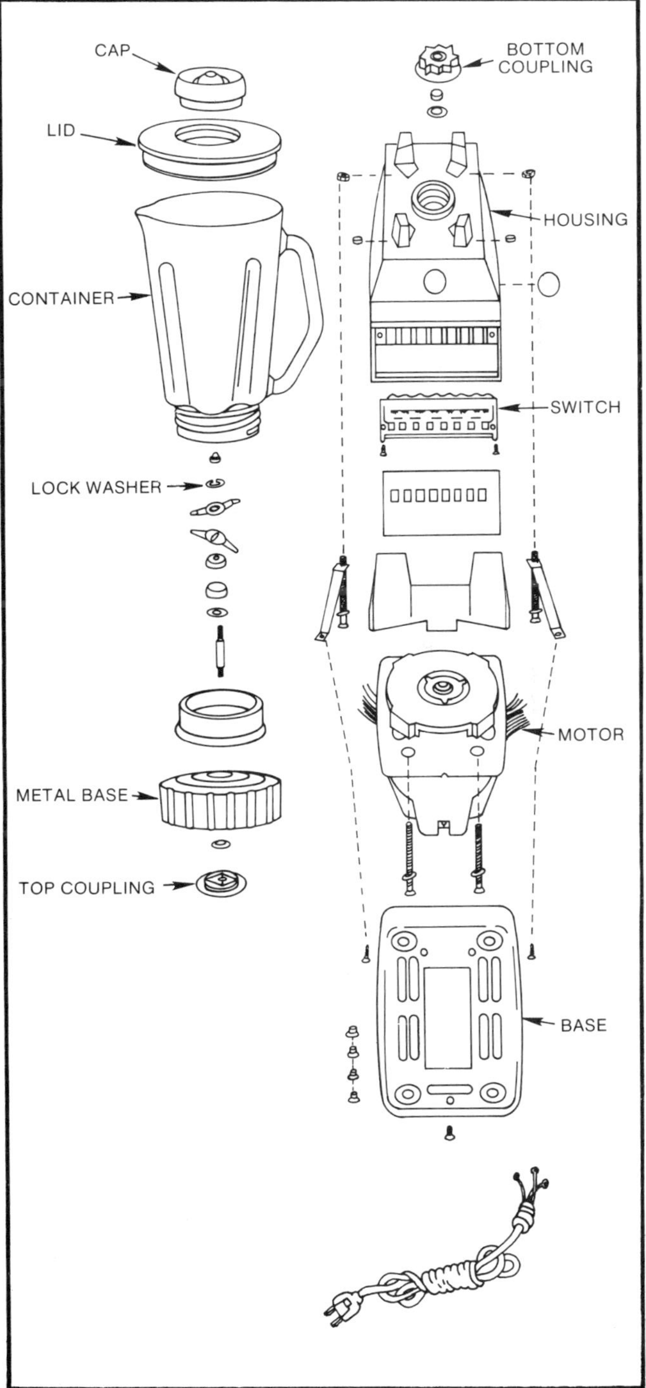

Exploded view of a typical blender.

All food processors operate similarly.

Food processors, which have become increasingly popular in recent years, are multi-purpose kitchen appliances that can perform a wide range of food preparation operations in a matter of seconds. Most food processors are capable of slicing; chopping; grating; shredding; mincing; crumbling bread, crackers, cookies, and all types of cereals; kneading bread and dough; pureeing; and mashing. A heavy, serrated cutting blade is used for grinding meat and kneading bread; a slicing disk slices vegetables and fruits into uniform pieces; a shredding disk shreds and grates; while a plastic mixing blade is used for whipping and blending. Actually, food processors are very similar to food blenders in terms of operation, with most problems common to both. The typical food processor motor operates on 115 V, 60 hertz, standard AC line current which generates a high torque and is capable of very high rotational speeds. But like blenders, most food processors have more than one speed available. Food processors operate in the same general power range as blenders, with 375 watts about average.

The operation of a food processor is simple enough. A food pusher is used to push food down a food chute. The food is pushed into a center bowl, where the appropriate blade or disk attachment is fit onto a center post (like the blender, this is simply an extension of the motor rotor and shaft). The rotating blade or disk then processes the food. Generally, the cover to the bowl has an interlock system that ensures that the processor will not operate unless the cover and interlock are in place. The bowl fits over the center post and is attached to the processor base by means of special side locking tabs. Besides housing the motor, the base usually is equipped with at least two control buttons or switches: one a continuous ON-OFF switch for long processing jobs, the other a pulse switch for short spurts of action.

There are few specific problems related to the operation of a food processor that are different from those encountered with a blender. In fact, food processors require a minimum of maintenance; however, you should make a point of wiping off any spillage from the base after each operation. The base can be safely cleaned with a mild liquid detergent, but it should not, of course, be immersed in water. If the motor in your food processor suddenly stops after you have been using it to work a thick or heavy mixture, the most probable cause is an automatic, temperature-operated circuit breaker that protects the motor. If the motor starts to get too hot, the circuit breaker, located in the base, will automatically cut off the current and the motor will stop. Turn the switch and the cover to the OFF position; then wait a few minutes and try to start the motor again. If it starts right away, there is no reason to suspect that anything is wrong with the motor.

Many times complaints are heard that a food processor did not do such and such a job adequately. Each model of food processor has different capabilities, so it is a good idea to read the owner's manual and find out what your processor can and cannot do well. Generally, food processors are not designed for beating egg whites to a large volume, whipping cream to a thick and fluffy consistency, finely slicing or shredding soft cheeses, mashing potatoes, slicing or shredding candied or dried fruits, and slicing hard-boiled eggs or hard nuts. Never try to process cheese which is so hard that you have trouble cutting it by hand, solidly frozen meat, or bones, and do not grind spices alone if they have a high oil content—cloves, for example.

Food Blender and Food Processor Troubleshooting Chart

Trouble	Probable Cause	Remedy
Motor does not operate in manual operating mode at any speed setting.	Operator has made an error, if there is a timer on the appliance.	Place timer selector to manual position. Set the desired speed and turn on.
	Line cord is defective.	Check line cord for continuity; replace if necessary.
	ON-OFF or hold-on switch is defective.	Check switch for continuity; replace if necessary.
	Motor is defective.	Check motor; especially check the armature and the field windings with your VOM for open or shorted wires. Replace parts as needed.
Motor with timer will not shut off at end of time selected.	Timer is defective.	Replace timer.
Motor will not shut off.	Hold-on switch is short circuited.	Check switch; replace if necessary.
Motor operates in manual, but does not operate in automatic.	Wiring connection is open.	Check wires and connections.
	Timer is defective.	Check timer; replace if necessary.
Motor operates, but not in all speeds.	Switch or switch wiring is defective.	Check switch and switch wiring; repair or replace as needed.
	Part of tapped-field coil, diode or speed control rectifier, or solid-state circuit is defective.	Test tapped-field coils for continuity. Remember, zero ohms indicates a shorted winding on a malfunctioning coil, while a reading of infinity indicates an open winding.
		Check diode or speed control rectifier with VOM.
		Check all wiring and switches to solid-state speed control circuit. After eliminating all other possibilities, replace the solid-state circuit.
Blender is noisy.	Rotor blade or shaft is bent.	Straighten, if possible, or replace.
	Motor bearings or blade shaft needs cleaning and lubrication.	Clean with alcohol and a toothbrush. Do not immerse the motor or control panel in water. Lubricate as recommended by the manufacturer.
	Motor armature is dirty or bent.	Remove armature from stator and clean. If armature is bent, it will probably have to be replaced. If the armature fan blades are bent, try to straighten them, or replace as needed.
	Loose screws, nuts, and other hardware.	Tighten hardware.

Food Blender and Food Processor Troubleshooting Chart (Continued)

Trouble	Probable Cause	Remedy
Container is excessively noisy.	Container bushing improperly positioned.	Reposition container bushing; if this does not quiet container, replace the universal coupling. Make certain the hole in the container is not off-center.
Container leaks.	Poor seal between container and bushing assembly.	Replace neoprene rubber seals or add an extra seal if necessary; also, tighten nut at the bottom of container.
	Loose fit between spindle and bearings.	Check and replace necessary parts.
Blender blows fuse or trips circuit breaker.	Short circuit.	Check line cord, switch, and motor for short circuits; replace defective part.
Case shocks user.	Short circuit to case.	Make continuity or ground tests to locate short circuit. Replace bare wires and tape terminals.

Electric Shavers and Hair Clippers

Electric shavers and hair clippers are relatively simple appliances used to remove and shorten hair. They vary in shape, size, and color, but their basic operational principles are the same. Shavers and clippers are generally driven by one of four types of power: vibrator, universal motor, contact-point motor, or for cordless models, a rechargeable battery.

Most of the early models of electrical shavers, as well as many of the current hair clippers, are of the vibrating-motor type which operates from the 60-hertz electromagnetic field of a coil. In this type of motor, an alternating current from the house line passes through an electromagnet, near the ends of which is suspended an iron bar called a vibrating armature. As the alternating current varies in direction and strength, it attracts the vibrating armature and repels it in rhythm with these variations, generally 3,600 per minute. The vibrating armature is connected to a set of small cutting blades, generally shaped like a comb, which are enmeshed with a stationary outer blade. Therefore, as the head of the shaver is pressed against the hair, it snips off the hair when it gets trapped between the fixed and movable cutting blades. The actual movement of the movable blades is very small.

The universal motor may drive a gear that rotates a number of rotary cutting blades, or it may drive a cam and connecting rod that transforms the rotary motion of a motor into a reciprocating motion. These high-speed oscillations once again drive the movable set of blades past the stationary one. For many years shavers used shaded-pole motors; however, today most use small, high-speed universal motors. The contact-point motor uses a coil that is controlled through a set of contacts linked to the armature. When the armature moves in one direction, the contact is broken and the armature begins to move in the opposite direction. As the armature moves, another contact is closed, drawing the contact until it reaches a point at which the contact is broken, sending the armature back again, and so on.

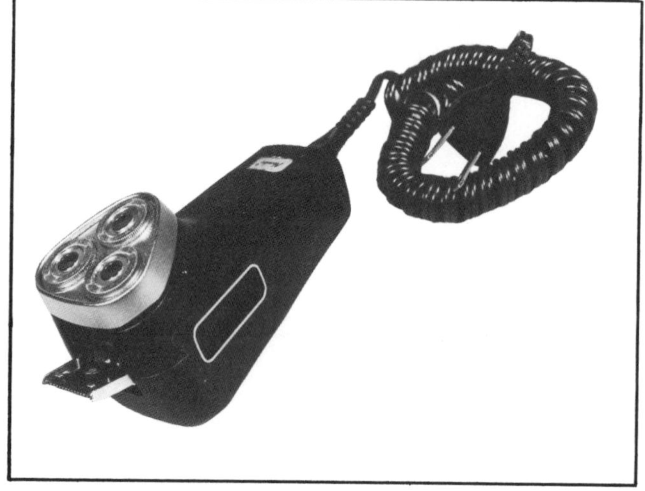

A typical rotary head electric shaver.

Portable cordless electric shavers usually employ a low-voltage—about 25 volts—universal motor which operates from a set of rechargeable batteries built into the shaver unit. There is also a recharger

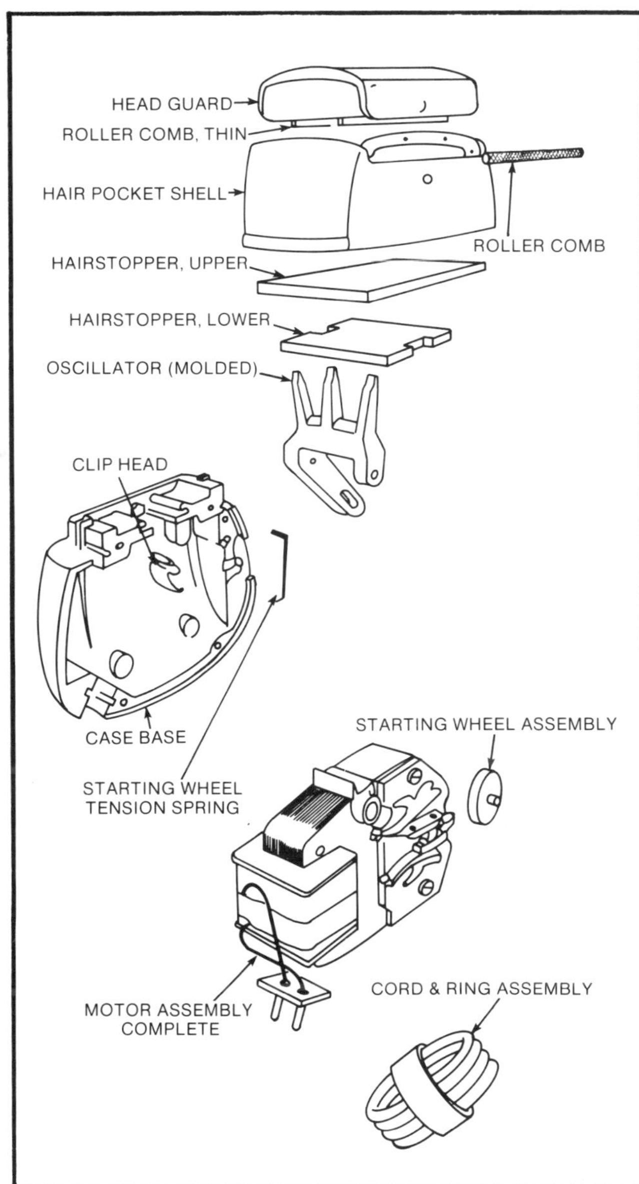

Exploded view of a typical cord model electric shaver.

which is used to keep the shaver's batteries charged.

Some shavers have dual-voltage motors so that the units may be operated on either the house convenience line or from an automobile power supply.

Actually, there is not much to a shaver or a hair clipper electrically. Mechanically, the motor bearings may become too tight; the armature may break or jam; or the transmission, shaver or clipper head, or blade assembly may break or become jammed. For the most part, you will need to replace parts, since repairing them is usually impossible. Make a habit of cleaning whiskers and hair from the head of the shaver or clipper with a fine, soft brush. Special brushes are available for this purpose. Lubricate the moving blade surfaces with electric-shaver lubricant or light oil, but be very sparing with the oil.

Electric Toothbrushes

Electric toothbrushes operate in a manner very similar to electric knives. Both appliances employ a small motor that converts electric energy into mechanical motion. Both use an eccentric transmission or a cam arrangement to change this mechanical rotary motion into the desired kind of oscillations. Most electrical toothbrushes permit the user to select either a back-and-forth or up-and-down brush motion. Although a few models have a cord arrangement, the vast majority are of the cordless type. The assembly of a cordless toothbrush, like that of a cordless knife, consists of two parts, the power handle and the battery recharger base. The handle must be stored in the well of the recharger, and the recharger must be connected to a continuously energized outlet to maintain a full battery charge. Take note that in some bathrooms, fixtures are so wired that the outlet is controlled by the light switch, which will result in an inadequate battery charge and either poor or no operation of the toothbrush. When the toothbrush is placed in the recharger, contacting terminals in the base mate with contacts on the toothbrush; the charger provides a small current trickle charge to continually charge the battery pack. A typical battery-powered toothbrush usually has a battery pack of five 1.2-volt DC rechargeable nickel-cadmium batteries. When the switch is closed, the battery powers the permanent-magnet motor, which drives a wobbler-gear device, causing the brush to oscillate.

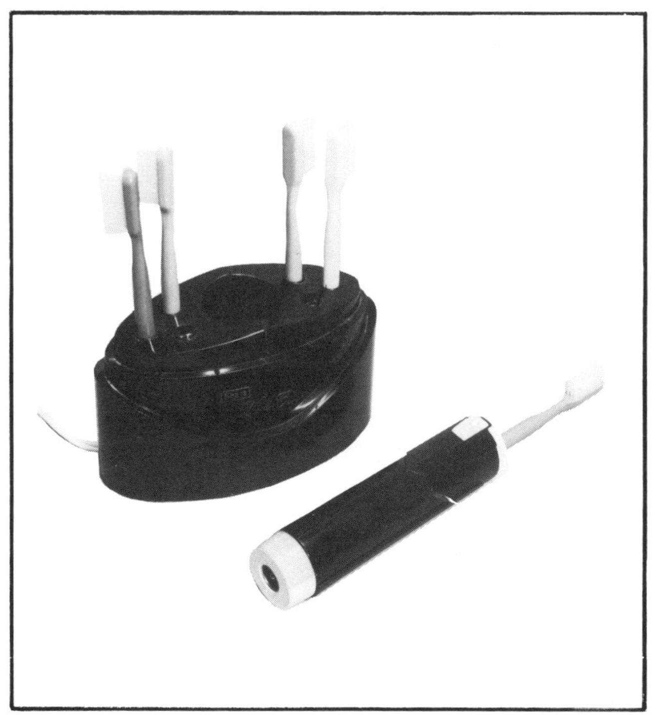

A typical cordless electric toothbrush with a battery recharger; electric toothbrushes also come in cord models.

Electric Shaver and Hair Clipper Troubleshooting Chart

Trouble	Probable Cause	Remedy
Motor does not run.	Line cord is defective.	Check continuity in line cord. Replace if necessary.
	Switch is dirty or defective.	Clean contacts. Check continuity in switch. Replace if necessary.
	Motor is defective; coil is open.	Test motor.
	Motor misadjusted (vibrator type).	Check and adjust air gap between rotor and stator according to manufacturer's instructions.
	Broken or worn motor brushes (motor-driven type).	Replace.
	Battery undercharged or dead (cordless type).	Charge battery or replace as necessary.
Motor runs, but blades do not move.	Vibrating armature assembly broken or jammed (vibrator type).	Correct; or if armature is defective, replace it.
	Shaver head dented or broken.	Replace.
	Partially shorted stator winding (motor-driven type).	Replace motor if necessary.
	Broken gear train, cam and follower, or similar mechanism.	Inspect. Replace faulty part.
Shaver gives poor shave, or clippers do not cut well.	Blades are dirty.	Clean.
	Head is dull or damaged.	Replace.
Shaver or clipper is noisy.	Loose screws or mechanical parts.	Tighten.
	Shaver head is bent or dented.	Replace.
Shaver or clipper runs slowly or erratically.	Loose wires.	Tighten and resolder.
	Dirty shaver head or switch.	Clean and oil.
	Dirty, burnt, or worn contact points (motor-driven type).	Replace contact points.
	Defective cell (cordless type).	Replace cell.
	Defective rectifier (cordless type).	Replace rectifier.

Maintenance is a key to keeping your electric toothbrush in tip-top shape. The power handle should be thoroughly rinsed under running water to remove toothpaste accumulation. The recharger base should be periodically disconnected from the power outlet and wiped clean with a damp cloth. Heavy accumulation of toothpaste either on the handle or in the recharger well will result in a poor seating of the handle and prevent proper charging. Also, before troubleshooting either the charger or the toothbrush, make sure that the spring contacts between the base and the toothbrush are clean and tight. Corroded terminals should be cleaned with sandpaper; be sure to remove all corrosion. If the spring contacts do not mate tightly, bend them out to make better contact.

Unfortunately for the home repairman/woman, most power handles and rechargers used with toothbrushes are hermetically sealed. This is primarily for watertightness, but it makes them nonrepairable. The cases are plastic and are cemented together so that they cannot be disassembled. Therefore, the troubleshooting procedure for most electrical cordless toothbrushes is limited to just determining whether the power handle or the recharger base is bad. To test the recharger, plug it into an energized 120-volt outlet. Insert a steel screwdriver into the charger well and up against the center metal post. A magnetic vibration should occur. If there is no vibration, replace the recharger.

Because of the induction system used in most electrical toothbrushes, the battery condition cannot be

Electric Toothbrush Troubleshooting Chart

Trouble	Probable Cause	Remedy
Handle does not run, runs slowly, or produces inadequate power.	Battery not charged.	Test recharger; if defective, replace. If recharger is all right, recharge battery.
		If recharger is not charging because contacts or terminals are dirty or corroded, clean them and then recharge battery.
	Outlet not energized.	Use another outlet which is continuously energized.
Handle is excessively noisy.	Main shaft is loose.	Check main shaft for looseness. A slight side-to-side movement is normal; but excessive movement indicates a defective assembly, in which case the handle must be replaced.
Toothbrush runs with normal speed but stalls when brushing.	Gear mesh is defective.	Check by turning toothbrush on, then grasping brush and stalling it. If the motor also stalls, check for a defective battery recharger; but if the motor continues to run, the gear mesh is defective and the handle must be replaced.
Case seam of handle is open.		Replace handle.
Recharger overheats.	Defective coil.	Check by touching center post in recharger; it should be warm to the touch but not hot. If it is excessively hot, the coil is defective and the recharger must be replaced.
Recharger hums.	Poor mechanical/magnetic circuit.	Replace recharger.

measured directly. However, you can determine whether the handle is defective and should be replaced or the battery merely needs to be recharged. If the handle does not run, operate the switch several times to be certain it is indexing properly. Turn the switch to OFF in models with an ON-OFF switch. Place the handle in a recharger unit known to be good. After 1 minute, turn the switch on, or press the push-push type once. If you see no motion of the shaft of a model with an ON-OFF switch, the handle is bad and should be replaced. The push-push type handle, however, should be tested further. Again press the switch firmly—once. Leave the handle in the recharger for another minute. If you see no motion of the shaft, the handle is bad and should be replaced. If motion is detected and it is sufficient to operate the shaft at least once, turn the switch off and leave the handle in the recharger for an extended period before deciding on its condition. An overnight charge should result in full capacity, power, and speed.

Vacuum Cleaners

All vacuum cleaners, irrespective of type, operate on the same basic principles and contain the same basic parts: a line cord to supply the necessary electric power, a switch of some type to control the operation of the unit, a motor-driven fan to provide the needed air suction, a duct system to channel the airflow, a nozzle to pick up the dust and dirt, and a case to keep the components together. While there are many different vacuum cleaners on the market today, all can be divided into three basic categories: upright, tank/canister, or combination. In addition, shop vacuums are heavy-duty variations of the tank/canister type.

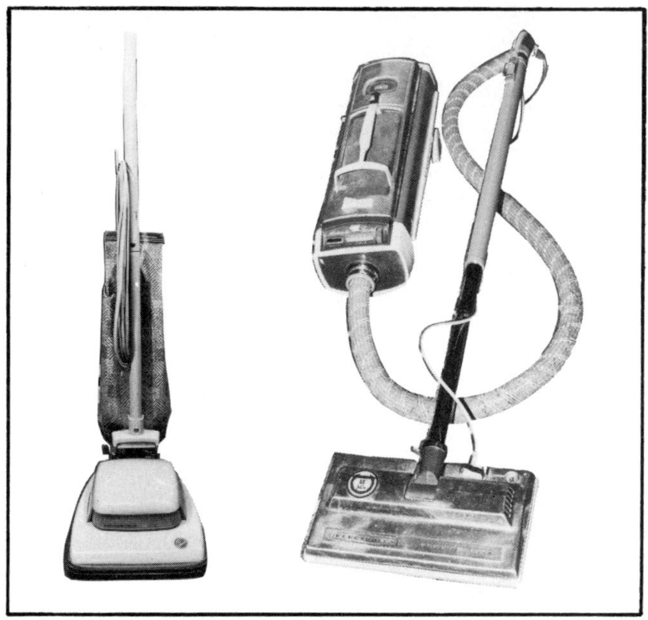

Two types of vacuum cleaners that are available: (left) upright and (right) combination.

The upright vacuum agitates the nap of a rug or carpet with a motor-driven brush to help to loosen dirt embedded in the pile and allow it to be sucked up into the bag. This brush also helps in picking up lint, threads, hairs, and so on. The motor, usually of the universal type, is connected directly to the suction-fan assembly and is connected by means of a belt-and-pulley arrangement to the revolving rotary-type brush. A fork handle is attached to the cleaner and serves to guide it over the carpet surface. Wheels attached to the casting incorporate an adjustment screw by means of which they may be raised or lowered for cleaning rugs of different thicknesses. In the typical upright unit, turning the switch ON completes the circuit and energizes the motor. The pulley mounted on the end of the armature drives the belt which turns the brush. The rotating brush loosens the dirt particles and tends to pick them up with centrifugal force. Airflow created by the fan of the armature pulls the dirt through the opening, around the brush, and through the motor base to a porous cloth bag. The air passes through the bag, depositing the dirt in it. The porous cloth bag must be periodically emptied out and replaced on the cleaner. Many units have a HIGH-LOW speed switch which regulates the motor speed and also mechanically closes off the intake from the bottom of the unit when in the HIGH position. This switch also lifts the brush free from the floor surface when in the HIGH position. Upright vacuum cleaners are excellent for sweeping rugs; their disadvantage is that the motor, base, handle, and dust-collection bag are all in one assembly, which must be moved back and forth across the room with each cleaning stroke.

Tank/canister models rely on suction alone without the motor-driven brush. The area of the nozzle in contact with the rug can be smaller and, therefore, the effective suction is considerably greater. Brushes and combs have been built into these nozzles to aid in picking up threads, hairs, and other small particles.

In most tank/canister vacuum cleaners, the collection bag is located at the intake end of the machine. The dirt-laden air sucked up through the nozzle and the hose passes through the bag, where most of the dirt is separated from the air. The air then passes through the motor and out the other end, where there is usually an additional dust filter to clean the air before it reenters the room. When a cleaner is plugged into a 120-volt AC power supply and the switch is pushed on, the motor circuit is energized. The armature with fans mounted on the shaft rotates in a counterclockwise direction—looking at the motor from the top. The rotation of the fans creates the airflow that starts at the swivel cap and goes through the disposable bag and inner bag, which filters out the dirt and dust. The air, after being filtered, continues to flow through the motor housing and fans and out the exhaust hole in the casing. Meanwhile, the dust and dirt collect in the disposable bag, which is discarded and replaced when it is about two-thirds full. The tank/canister type of vacuum cleaner has the advantage of great versatility; with a flexible hose and a variety of nozzles, it can be used to clean in tight corners, along baseboards and heating elements, to pull cobwebs down from the ceiling, to vacuum the interior of an automobile, as well as to do standard house vacuuming. On some models the hose can be connected to the air outlet to create a stream of air pressure. The tank/canister type of vacuum cleaner is not, however, nearly as good as the upright model for heavy-duty rug cleaning.

In the shop vacuum, air and dirt are drawn up through the hose; dust and dirt drop into the container as the air passes through a porous paper filter bag and an inner porous cloth bag. Very fine dust clings to the outer porous bag and as it accumulates, it decreases the efficiency of the vacuum cleaner. Instead of a porous outer bag, some shop vacuums utilize a permanent paper-type washable cartridge filter.

Some manufacturers combine the strong suction of the tank/canister model and the powered agitator head of the upright. They do this by adding a powered rug nozzle with a separate motor to the tank/canister model. Double duty is the feature of a combination unit. The canister, with attachments, offers the powerful suction that is important for general floor care, above-the-floor care, and the versatility that is unique to the canister. The powered

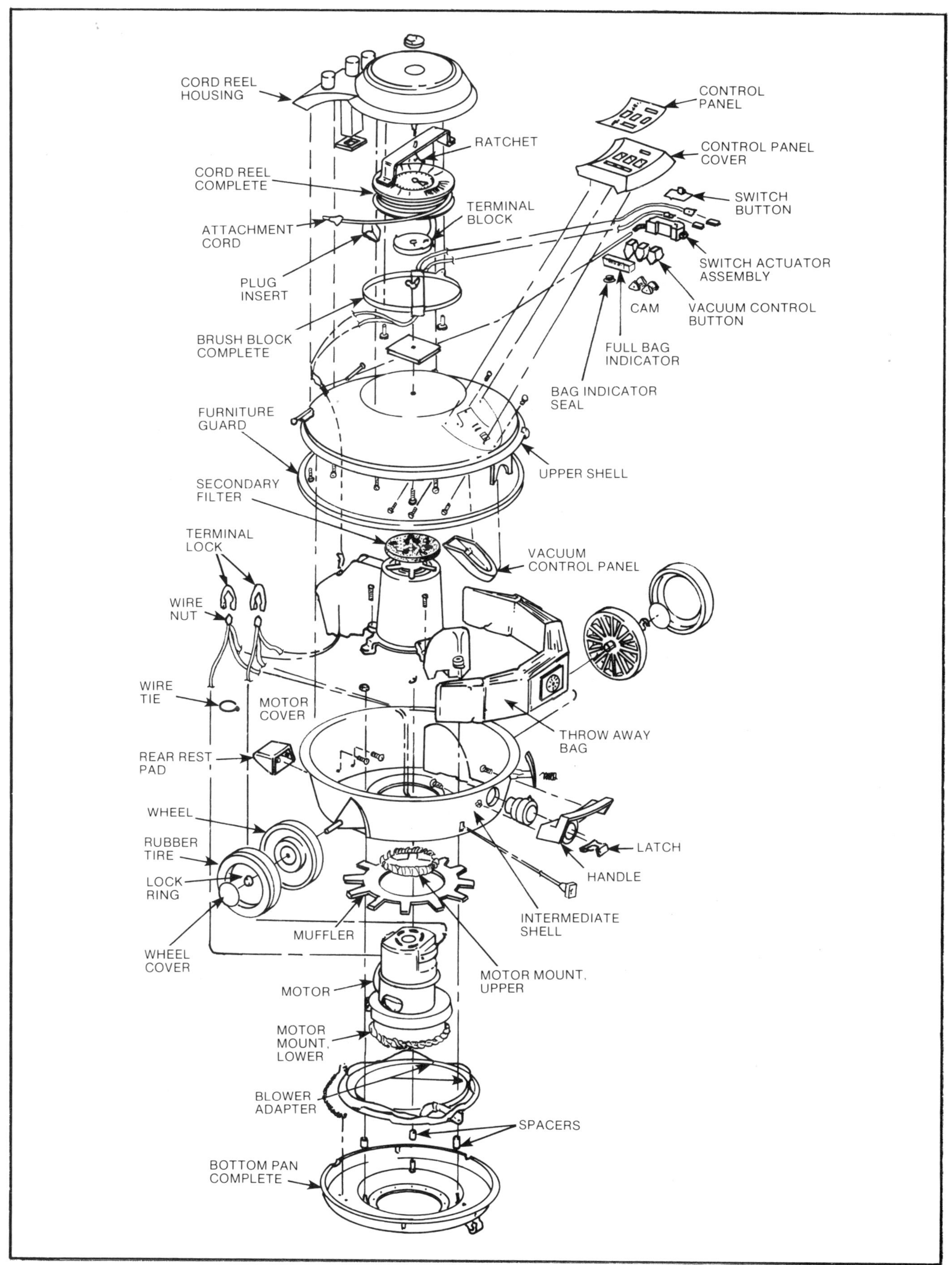

Exploded view of a typical canister vacuum cleaner.

nozzle attachment with its separate motor-driven brush agitates the rug and provides rug-cleaning efficiency. In this way, the combination unit is similar to the upright, but it is easier to operate.

Any vacuum cleaner loses its suction power and, therefore, its ability to remove dust and dirt when the porosity of the bag is inhibited. If air cannot pass through the bag, cleaning ability is hampered. Fine particles of dust collect very easily on the bag, especially on the bags of shop vacuums. It is essential to remove the caked dirt by either dumping the bag of upright vacuum cleaners or replacing the bags of tank/canister models. Clean bags allow the flow of air to develop the proper suction, and the motor is protected from the dust.

One of the most often-heard complaints regarding a vacuum cleaner is that it doesn't clean as well as when it was new. In upright vacuums, this problem can be attributed to one or more of the following:

1. The bag is caked with dirt. If provided, remove the inner bag and replace it with a new one. Clean the cloth bag.
2. Check the fan blade; it may be slipping on the motor shaft. Tighten the fan-blade setscrew, if applicable.
3. Check the tightness of the drive belt which turns the agitator. If it is loose and slipping, tighten it. If the motor is replaced, check the belt tracking. Align the motor so that the belt tracks correctly; then secure the mounting hardware.
4. Inspect the agitator brushes; if the bristles have become shortened because of wear, adjust the position of the agitator to move it closer to the floor. If the agitator is not adjustable or if the bristles are too short, replace the agitator.

In tank/canister models, the problem of little or no suction is usually a plug of dirt stuck in the accessory, the hose, or the air inlet. Often, a toothpick or hairpin is caught sideways; dirt and hair then begin to cling around the obstruction, causing more of a problem. A decrease in suction often takes

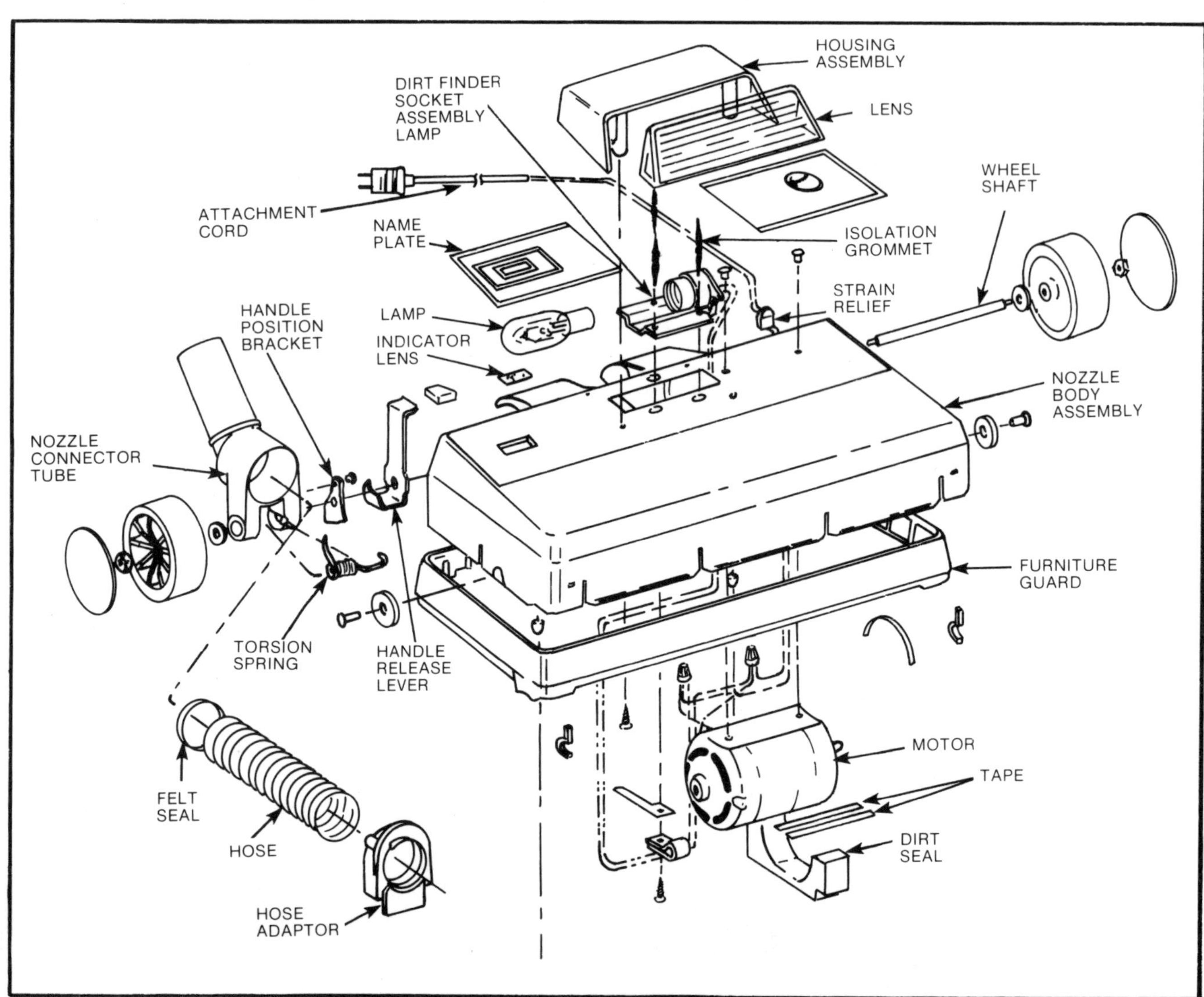

Exploded view of the powered nozzle used with a combination vacuum cleaner.

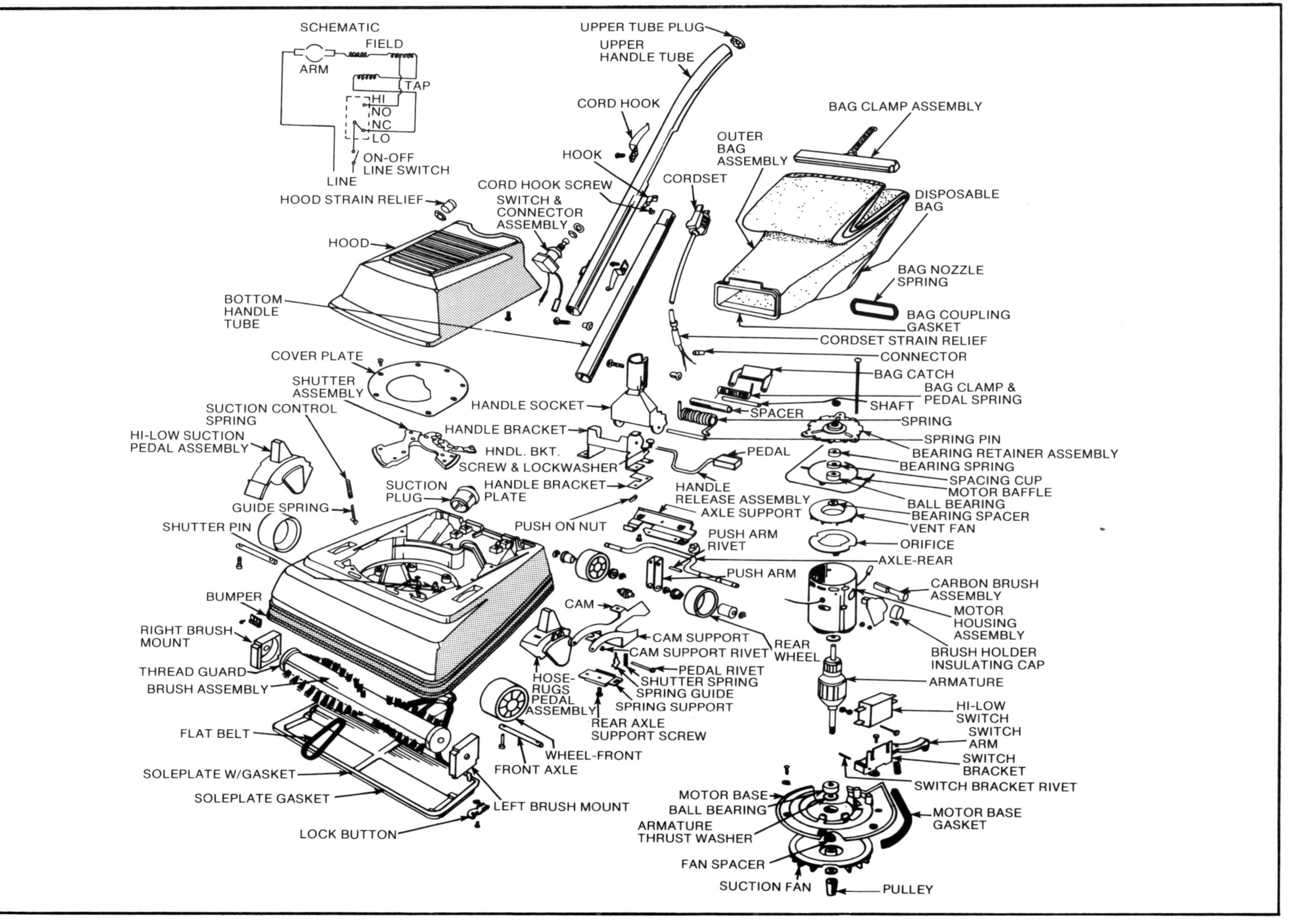

Exploded view of a typical upright model vacuum cleaner.

place slowly over a period of time, and the user will not usually realize it is happening. Simple dirt plugs can be cleared by quickly closing and opening the end of the hose with the palm of your hand. If this does not dislodge the dirt, remove the hose from the air inlet and connect it to the air outlet if your vacuum is so adapted; sometimes this force of air in the opposite direction can dislodge the plug. If all else fails, run a heavy wire into the hose to dislodge the plug.

Other problems in tank/canister or combination vacuum cleaners that prevent them from operating the same as when they were new are:

1. The hose and accessory couplings are cracked or chipped so that air leaks in; this decreases the sucking pressure at the nozzle. Replace the parts as required.
2. The gasket between the fold-back top or fold-out end and the vacuum-cleaner base leaks air; this also decreases the suction at the nozzle. Replace the gasket, if required.
3. The motor-driven brush is not rotating on combination models. Check the motor and drive belt. The small drive belt is prone to slip off; if that is the case, you just have to put it back in place, unless it is clearly too loose. If the belt is too loose or broken, replace it. If the motor is replaced, check the tracking of the drive belt. Align the motor for proper tracking and then tighten the motor mounting hardware.
4. The brush bristles on the agitator are worn on combination models. Adjust the position of the agitator to move it closer to the floor. If the agitator is not adjustable or if the bristles are too short, replace the agitator.
5. The blower assembly is loose on the motor shaft; tighten the setscrew.

Noise in a vacuum cleaner may be caused by a loose blower assembly, by broken blower fins, or by fins that are out of position and are striking something; tighten, realign, or replace as required. Constant vibration may cause nuts, bolts, or machine screws to loosen; inspect and tighten any loose hardware. Finally, the motor bearings may need lubrication or may be defective; lubricate or replace as required.

Vacuum Cleaner Troubleshooting Chart

Trouble	Probable Cause	Remedy
Motor will not run.	Line cord is defective.	Check and replace line cord.
	ON-OFF switch is defective.	Check and replace ON-OFF switch.
	Worn motor brushes or brushes hanging up in the brush holder.	Replace brushes or make adjustments.
	Jammed fan.	Remove cause of bind. If fan blade is bent or damaged, replace it.
	Frozen motor bearings.	Clean and lubricate; if too worn, replace bearings.
Motor overheats.	Motor is clogged with dirt.	Clean motor.
	Hose and/or cleaning tool and/or bag collar connector is plugged. Hose is not connected to cleaner inlet properly. Ventilation openings are filled or clogged with dirt. Bag is full.	All of these possibilities can cause a motor to run hot because they all restrict airflow to the motor. If a motor has run hot for a considerable time and damage has occurred, field coils and/or armature will usually be discolored.
	Motor is defective.	Check motor.
	Loose wire connections.	Tighten connections or replace components as necessary.
	Pores of throw-away filter bag are plugged. This can occur rather quickly if extremely fine dirt, such as flour, fireplace ash, and so on, is picked up in quantity or if the user attempts to reuse the bag.	Replace filter bag.

Vacuum Cleaner Troubleshooting Chart (Continued)

Trouble	Probable Cause	Remedy
Motor overheats.	Foreign material binding the brush assembly (upright and combination nozzle).	Remove the source of the bind.
Motor runs, but cleaner has little or no suction or will not pick up.	Hose and/or cleaning tool and/or bag collar connector is plugged.	Clean the plug out.
	Bag is full.	Install new bag.
	Hose is not connected to cleaner inlet properly.	Check for loose connector, improper or loose fit of handle.
	Secondary motor filter is clogged.	Clean filter.
	Seal between motor and shell is poor.	Check for misplaced gasket around motor.
		Tighten loose nuts which secure motor and bottom pan.
	Seal between top or end cover and cleaner body is poor.	Check for misplaced gasket around motor.
		Tighten loose nuts which secure motor and bottom pan.
	Seal between top cover and cleaner body is poor.	Check unit for damage, warping, or obstructions that would prevent a good fit.
		Check for misadjusted or damaged gasket.
	Fan is loose on motor shaft.	Replace fan as necessary and tighten fan nut securely.
	Improper installation of switch strain relief in motor cover.	Ensure proper installation.
	Air leaks in hose or other attachments or in body of vacuum.	Check for air leaks and replace parts as needed.
	Defective nozzle adjustment mechanism or handle-tension spring (upright model).	Check and replace as necessary, or make necessary adjustments.
	Damaged or misadjusted agitator or floor brushes (upright or combination nozzle).	Adjust bristles flush with the bottom casting lip or lower; otherwise replace.
	Incorrect nozzle adjustment for carpet nap (upright or combination nozzle).	Adjust nozzle for correct contact.
	Broken agitator belt (upright or combination nozzle).	Replace belt.
Cleaner runs, but power nozzle will not run (combination model).	Nozzle cord is not firmly plugged into receptacle on vacuum hose.	Check connection.
	Vacuum hose cord is not firmly plugged into receptacle on vacuum body.	Check connection.
	Open circuit in wiring.	Check for continuity on the vacuum receptacle and wiring, vacuum hose wires, and the power-nozzle attachment cord.

Vacuum Cleaner Troubleshooting Chart (Continued)

Trouble	Probable Cause	Remedy
Cleaner runs, but power nozzle will not run (combination model).	Open circuit in motor.	Check motor brush to be sure it is not hanging up.
		Blow out motor with air.
Operation noisy or excess vibration.	Motor fan is loose, bent, or broken.	Tighten or replace fan.
	Loose rivets, screws, or other hardware.	Tighten as necessary.
	Motor is not securely mounted.	Motor should be properly installed and nuts that secure bottom pan should be tightened.
	Blower assembly is loose or fins are broken.	Check and replace as necessary.
	Motor bearings need lubrication.	Lubricate bearings.
Cord reel will not rewind.	Defective cord jamming reel.	Remove bind.
	Broken cord reel spring.	Replace spring.
	Related parts are loose.	Tighten as needed.
Bags bursting.	Picking up large amounts of fine dirt such as fireplace ash, flour, baking soda, and so on.	Change bag frequently when picking up fine dirt.
	Attempting to empty and reuse throw-away bags.	A new bag should be used each time a bag needs changing.
Dust leaks into room.	Holes in dust bag.	Replace dust bag.
	Cloth bag extremely dirty.	Clean or replace if necessary.
	Dust bag improperly installed.	Check owner's manual for proper installation.
	Defective sealing gasket.	Replace gasket.
Vacuum cleaner blows fuse or trips circuit breaker.	Short circuit.	Make continuity tests on all vacuum cleaner electrical parts, until short circuit is found.
Vacuum shocks user.	Short circuit to case.	Test for short circuits. Replace bare wires and tape terminals.

Fans

Electric fans are one of the simplest types of motor-driven appliances, having only a few primary components. The function of these components is to circulate air. This commonly serves a cooling function, but fans are also used to circulate warm air or to serve an exhaust function—drawing odors out of kitchens, hot, moist air out of bathrooms, and hot air out of attics. Some fans used to circulate air in a room are in fixed stands that can be tilted in a variety of positions, from vertical to horizontal, to direct the flow of air. Other fans, called oscillating fans, have internal gearing mechanisms that cause them to swivel back and forth, changing the direction of the circulating air. Fans are usually mounted and have a round or square guard around them for protection. A fan consists of a motor with an assembly of two to five blades attached to the motor shaft. The spinning motor rotor shaft turns the blade assembly, which circulates the air. Oscillating fans have a mechanical system that is also driven by the motor shaft, causing the fan to swivel back and forth. Most fans have an ON-OFF switch, and some fans have speed controls, direction switches for drawing air in or blowing air out, and thermostats to control the ON-OFF condition of the fan depending upon the room temperature.

Fan motors are usually shaded-pole or split-phase induction motors having solid rotors; therefore,

A typical home-model electric fan.

they are practically maintenance-free. The motors are usually less than ¼ horsepower. Since they are shaded-pole or split-phase induction motors, the starting and running torques are low, which is an added safety feature should children's fingers or other objects be placed in the path of a rotating blade. Some motors are reversible by means of a circuit that short-circuits one of two shaded-pole turns. Fan blades are made of light metal or plastic—the larger the diameter, the greater the pitch; and the greater the length of the blades, the larger the volume of air circulated. Of course, the more air the blades are capable of moving, the more powerful the motor must be. The blade assembly is held to the motor shaft with a setscrew or is friction-fit by means of a compression ring. To remove a friction-fit blade, carefully pry the assembly off; use a rag to protect the fan frame from damage. There may be a slot in the rotor shaft to mate with a web in the fan blade hub. When replacing the blade hub, be sure to engage the web in the slot and place the hub all the way on the shaft. The web and slot prevent the blade assembly from slipping on the shaft during operation.

Oscillating fans have a gear mechanism that causes the fan to swivel back and forth. A worm gear on the motor shaft drives a pinion. A secondary gear and clutch disk operate to drive the vertical shaft and crank. The crank attaches to the oscillating plate, which causes the fan to swivel in the swivel part.

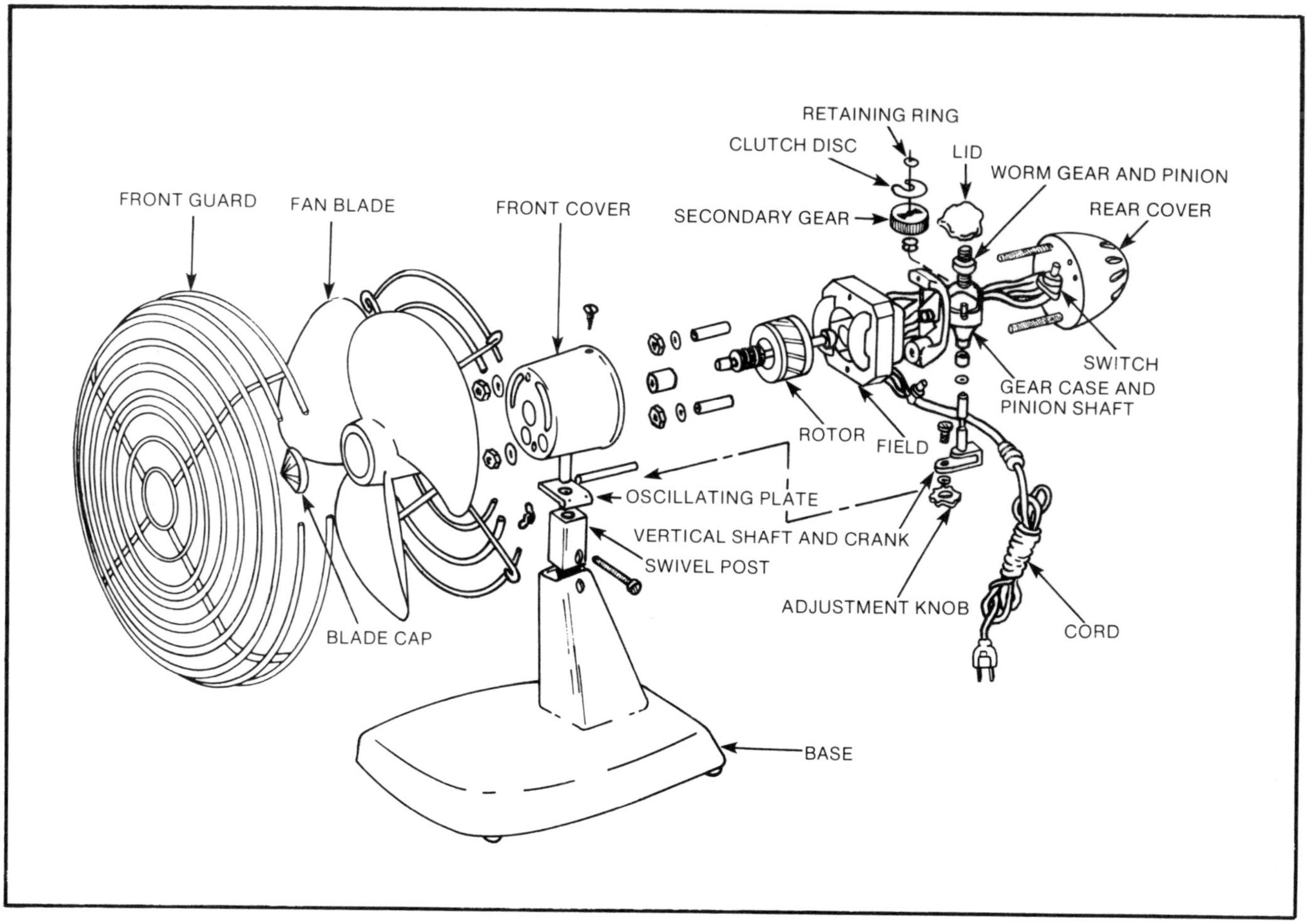

Exploded view of a typical oscillating fan.

Fan circuits are simple electrically and can be tested by making continuity tests, making sure that the line cord is disconnected from the convenience outlet. Electric current from the convenience outlet flows through the contacts of the adjustable thermostat to the ON-OFF speed control switch. When the switch is in the LOW position, current flows through a choke—an inductor—and the stator winding of the shaded-pole induction motor. The current flow through the inductor causes less current through the complete circuit and, therefore, the motor runs at a lower speed. When the speed control switch is set to HIGH, all the current flow is through the stator winding of the motor, causing it to run at a higher speed. If the temperature in the room drops below the setting of the thermostat, the contacts open, breaking the current flow to the motor. The motor rotation direction is altered by the positioning of the IN-OUT direction switch; the switch short-circuits one or two sets of shaded-pole windings on opposite ends of the stator winding laminated core.

Kitchen exhaust fan ON-OFF switches are spring-loaded with normally closed contacts. The switch is actuated by a chain that is positioned in such a manner that the chain holds the switch depressed when the fan is off and the outside cover door is closed. When the door is opened by releasing the chain, the force on the chain is removed from the switch, which springs out to its normal position; the electrical contacts close, permitting current flow to the motor.

Since fan motors are mostly induction motors without brushes or a commutator, motor failure is minimal. Mechanical problems and maintenance include excessive vibration, blades striking the frame, failure to oscillate, frozen bearings, lubrication, and accumulated dirt. Excessive vibration is caused by blades that are out of balance, usually because a blade has been hit and bent by some object. Metal blades can sometimes be straightened. To straighten blades, rotate the fan blade assembly by hand. Each blade should be a fixed distance away from an established reference point. Carefully bend the blade that is out of balance and check again with the reference point. If a blade cannot be straightened properly or if excessive vibration is caused by a broken or damaged synthetic blade, the complete blade assembly must be replaced. If a blade hits the frame of a fan, bend the frame away from the blades, then check the blades for bends as just described.

If there is a doubt as to whether a malfunction is in the motor or in the gearing system, isolate the problem by removing the motor from the gearing. Spin the motor rotor by hand. If it spins easily and coasts to a stop, the problem may be electrical, in the motor or in the gearing. Connect the motor to alternating current; if the motor runs, the problem is in the gearing.

If an oscillating fan fails to swivel, there is a problem with the gearing or crank mechanisms. Inspect the crank and oscillating plate first, because disassembly is not necessary. If the problem is not located, remove the rear cover and inspect the gearing. A stripped, worn, or broken gear, pinion, or motor shaft worm gear must be replaced; if one part is worn, its mating part is also usually worn and should be replaced. If no gears are broken, stripped, worn, or misaligned, the parts may simply be excessively dirty. Clean the parts in alcohol, using a toothbrush to remove stubborn dirt and old lubricant. Replace the gear grease in the gearbox. Lubricate other moving parts at pivot points with a coat of SAE-30 motor oil.

To determine if a fan needs lubrication, turn the blade by hand. If it turns freely and coasts to a stop, it does not need lubrication. If the blades turn sluggishly or do not coast, place one drop of light oil on each bearing through the oil hole. Run the motor up to speed to work the oil in. Wipe off any excess oil. If there are no oil holes, remove the end belts, oil the wicks, and reassemble.

If the fan blade is belt-driven from the motor—large attic fans, for example—check for proper tension. To check the tension, use your finger to depress the belt at a point halfway between the blade and the motor. The belt should deflect ¼ to ½ inch when set at the proper tension; too much tension places a sideward stress on the motor bearings. The pulley of the motor mount can be moved to adjust the tension. If the belt is worn, replace it.

The exterior surfaces of a fan—the blade guards, the blades, and the case—should be cleaned with a cloth dampened with a solution of detergent and water. Keep the solution out of the bearings and electrical circuits. Use an old toothbrush to loosen stubborn dirt.

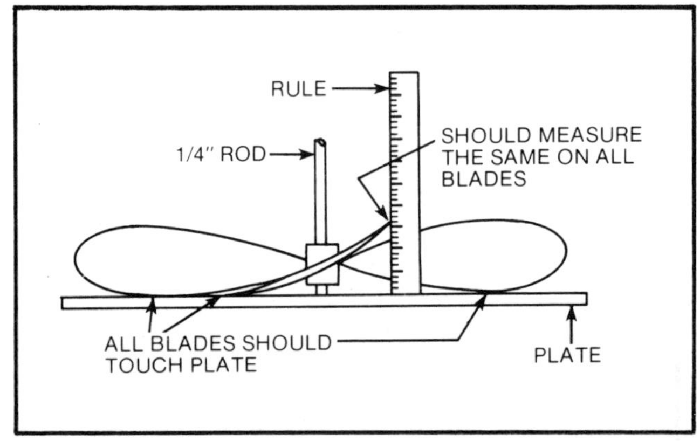

Method for checking a fan blade for proper pitch and tracking.

Fan Troubleshooting Chart

Trouble	Probable Cause	Remedy
Motor will not run.	No voltage to convenience outlet.	Check for voltage.
	ON-OFF switch is off or thermostat contacts are open.	Turn switch on and rotate thermostat to maximum.
	Line cord is defective.	Check line cord for continuity.
	ON-OFF switch is open.	Should have continuity with switch on, none with switch off.
	Thermostat contacts are open.	Test thermostat.
	Motor stator winding is defective.	Test motor for continuity with VOM on R x 1. If circuit is open, replace motor.
	Variable resistor open.	Check with VOM on R x 100.
Fan vibrates.	Blade is out of balance.	If blade is metal, set up a reference surface and try to straighten bent blade. If it still vibrates or if blade is synthetic material, replace blade.
	Loose or missing screws.	Replace as needed.
Blade strikes frame.	Frame is bent.	Straighten frame.
Fan is noisy or blades do not turn easily by hand.	Lubrication is insufficient.	Lubricate motor bearings. Lubricate crank and gear assemblies.
	Part is loose.	Inspect. Tighten loose parts.
	Motor bearings dry or worn.	Lubricate or replace.
Blades turn only at low speed.	Motor bearings are dry.	Lubricate.
	Blade is loose on motor shaft.	Tighten setscrew.
Blade assembly of belt-driven fan does not operate; slips while operating.	Belt is broken.	Replace belt.
	Belt is slipping.	Increase tension on belt.
Oscillating fan fails to oscillate.	Crank or gear mechanisms are broken, worn, or stripped.	Inspect crank mechanism.
		Inspect gear mechanism and replace damaged or worn parts.
Fan blows fuse or opens circuit breaker.	Short circuit.	Test line cord, including plug, switch, and motor. Replace defective part.
Case shocks user.	Short circuit to case.	Make ground test to locate short circuit. Replace bare wires and tape terminals.
Fan with reversible feature does not reverse.	Defective selector switch or switch connections.	Check for continuity and replace as necessary.
	Motor stator winding is defective.	Check for continuity with VOM; if faulty, replace motor.

Repairing Small Appliances That Heat

The heating effect of electricity makes possible many modern small appliances, including slow cookers, frying pans, coffee makers, toasters, hair setters and dryers, clothing irons, electric blankets, and electric heating pads. The heat developed in these appliances is obtained by passing current through a special wire or element. This element has a higher resistance to the passage of electricity than does ordinary wire. Overcoming this resistance causes heat; therefore, the process is known as resistance heating. For the most part, this group of appliances is among the simplest to repair. Generally, they have only four electrical sections that can malfunction: the electrical power source, the transmission line, the control, and the power-conversion section. In addition, the mechanical systems of many of these appliances are simple as well, although toasters and steam irons are rather complex. It stands to reason that many of the problems that are encountered with small motorized appliances will not occur with this group, if for no other reason than a lack or absence of moving parts. Electrically,every small heating appliance consists of at least a plug, a line cord, and a heating element. The more sophisticated types may have two or more heating elements, an ON-OFF switch, a multiposition heat-selector switch, a variable thermostat, and a safety thermostat. Some of the heating appliances, hair dryers and rotisseries, for example, also incorporate a motor that drives fan blades which in turn blow warm air or perform some other function related to the heat producing ability of the appliance. As with the small motorized appliances, we have grouped all types and variations of each appliance in a single group.

Slow Cookers

Slow cookers, crock pots, bean pots, country kettles, and fondue pots are used to cook food at a low temperature over a long period of time. For example, a stew or a recipe for beans may be placed in a slow cooker in the morning and the cooker turned on at a low temperature. The stew or bean dish cooks all day, and by evening the food is ready for eating. The fondue pot is used to bring cheese, chocolate, or a similar substance to a temperature above its melting point so that foods may be dipped into the fondue pot to apply the warm, melted dip to the food.

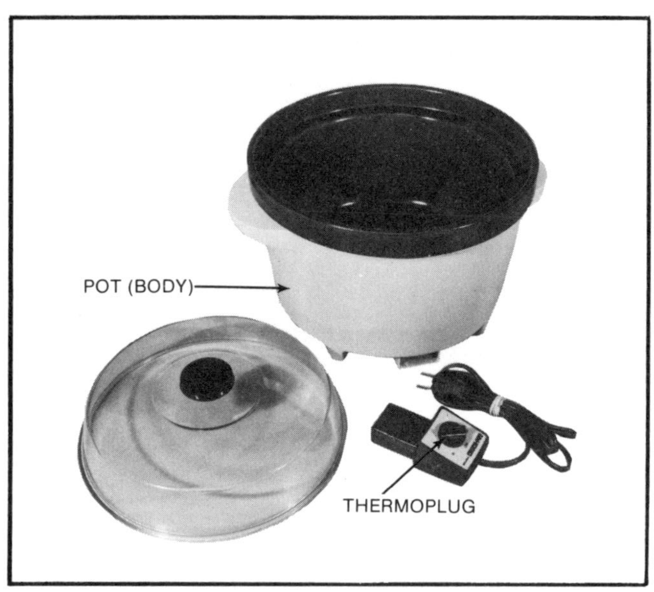

A typical crock pot with a thermoplug-type base.

The appliances in this category are made in two configurations. The typical slow cooker has a heating base upon which a ceramic pot sits; the heating base is separate from the pot, making the pot easy to clean. The heating base can be used separately for other utensils and the pots can be used on the stove instead of on the heating base, if desired. The heating base contains a heating element which is replaceable in some appliances as a complete unit and is not replaceable in others. The heating base also includes a thermostat and terminals which connect to the input power cord. The thermostat front-panel dial is numbered to indicate relative heat settings. In slow cookers, the thermostat is not calibrated to any specific temperature, and the dial settings do not normally correspond to any specific temperature. By advancing the thermostat control clockwise, the operating temperature of the heating base is increased. Although you have to get a feel for the relative heat of the settings, recipes that are usually supplied with a slow cooker advise the user where to set the thermostat.

The second common configuration for these appliances is typified by the construction of the country kettle. In this appliance, the heating element is molded into the base or body of the appliance. The terminals are hermetically sealed to prevent moisture from getting into the body of the pot and the heating element when it is immersed in water for

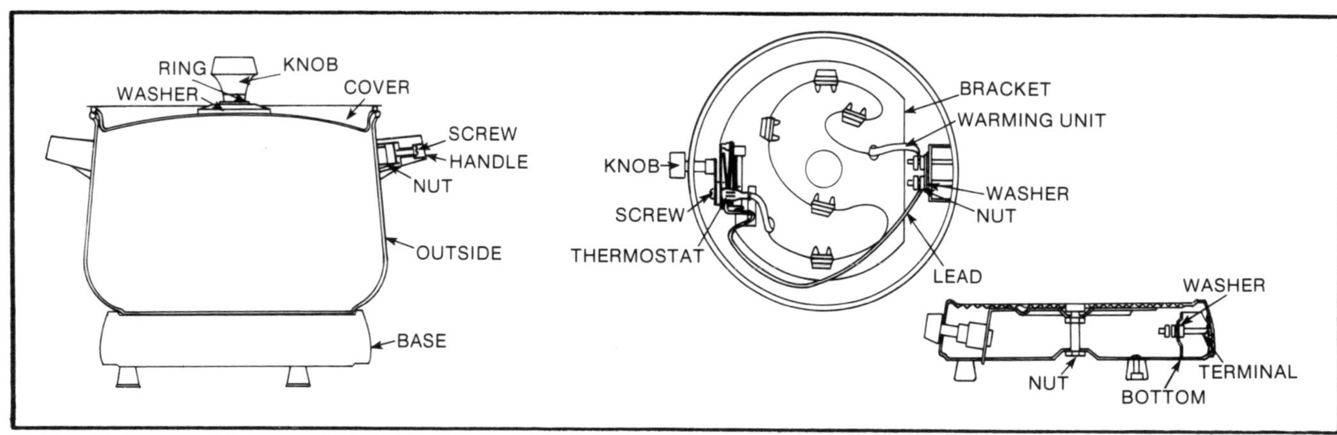

Another type of crock pot in which the pot and the heating base are separate.

Slow Cooker Type Appliances—Including Those with Thermoplugs—Troubleshooting Chart

Trouble	Probable Cause	Remedy
Cooker does not heat.	No input voltage.	Place ON-OFF switch on or set adjustable thermostat to maximum heat.
	Line cord is defective.	Check for continuity; replace if necessary.
	ON-OFF switch is defective.	Check for continuity; replace if necessary.
	Thermostat is defective.	Check for continuity; replace if necessary.
	Heating element is defective.	Check for continuity; replace if possible, or replace cooker base.
Cooker temperature is not maintaining a constant range; cooker gets too hot; cooker does not get hot enough; or indicator light does not cycle on and off.	Thermostat is defective.	Check to see if thermostat contacts open and close. Clean contacts. Repair or replace if necessary; then calibrate thermostat.
Indicator light does not come on.	Light and/or resistor is defective.	Check light and resistor for continuity; replace parts if necessary.
Cooker will not shut off.	Short-circuited ON-OFF switch or adjustable thermostat contacts shorted.	Check switch and thermostat for resistance; replace parts if necessary.
Cracked or damaged pot, kettle, or other container.		Replace broken part as soon as possible.
Cooker blows fuse or trips circuit breaker.	Short circuit.	First check line cord for resistance; then make a visual inspection of the heating element, if possible. Check heating element for resistance. Check for internal short circuits. Replace defective parts or replace cooker base, if heating element is defective on thermoplug model.
Cooker shocks user.	Short circuit to cooker body.	First check visually; then check for continuity. Replace defective parts or replace cooker base if heating element is defective on thermoplug model.

cleaning. The power cord assembly, called a thermoplug, contains a thermostat that causes the pot or base to maintain a constant temperature. The thermoplug usually consists of a variable thermostat, a neon indicator light, and connectors that mate with the body of the kettle providing power to the heating element. The thermoplug also contains a bimetallic strip in a probe that inserts into a hole in the body of the kettle. The probe detects the heat on the bottom of the body of the kettle and, in so doing, causes the thermostat bimetallic strip to bend back and forth in accordance with the temperature detected. The advantage of a thermoplug is that it can be used with a number of different appliances to provide power to and temperature control for the heating elements within the appliances.

When a slow cooker that contains an adjustable thermostat within the cooker body or base is connected to a convenience outlet, 120 volts AC is applied to the adjustable thermostat control, which also acts as an ON-OFF switch. When the thermostatic control is advanced clockwise, the appliance turns on and current flows through the thermostat to the heating element, which opposes the flow of current and, thus, produces heat to cook the food within the appliance. As mentioned before, this is known as resistance heating. When the bimetallic strip senses that the operating temperature as set by the user has been reached, the thermostat contacts open, interrupting the current to the heating element, which turns the appliance off momentarily. When the bimetallic strip senses that the appliance and food temperature have decreased approximately 5 to 10 degrees F, the thermostat again closes, allowing the flow of current to the heating element to heat the appliance. Slow cooker thermostat controls can be set to various temperatures by the user; however, the graduations on the cooker dial usually read from 0 to 5, 0 to 10, or Low heat to High heat and are not related to specific temperatures.

The only malfunctions that occur in this type of appliance are no heat, too much heat, or no heat control. In the case of no heat, the heating element is either open or is not receiving electrical power. First use your VOM to check for voltage in the line cord. Next, use the VOM to see if the thermostatic control is working; be sure to rotate the thermostat fully clockwise during this check. The absence of 120 V AC indicates that the thermostat is malfunctioning, and if a repair is not obvious, the thermostat must be replaced. Next, use your VOM to check for voltage in the heating element. If there is 120 V AC present on both sides of the element and the element is not on, there must be an open circuit in the heating element. Depending on the type and model of cooker, the element or the element unit may or may not be replaceable. If the element is not replaceable, the cost of a new appliance may not be much greater than the cost of the repair.

If the appliance is receiving too much heat, it indicates that the thermostat control is not turning off. This may be checked by using the VOM or by visually observing if the heating element goes on or off, if the heating element can be observed. No heat control can also mean that the thermostat is not operating properly; the contacts are probably closed and must be replaced.

The other basic type of cooker—one with a thermoplug—has an internal heating element that is not separable from the body of the appliance. If the heating element on this type of cooker is defective, the body must be replaced. On this type of appliance the thermoplug thermostat is adjustable and is used for turning the appliance on or off as well as selecting the operating temperature. If the dial is calibrated in degrees Fahrenheit, the markings indicate the correct temperatures for the cooker for which the thermoplug was purchased; however, the thermoplug may often be used with a number of different appliances or utensils, and therefore the operating temperatures indicated on the dial may not truly indicate the actual temperatures for all of the appliances used with it. The adjustable thermostat and the neon bulb with the current-limiting resistor are always integral parts of the thermoplug unit. If it is necessary to disassemble a thermoplug for testing and repair, you may need to punch a small hole in the center of the temperature dial nameplate to get to a screw that holds the dial in place. Insert a tiny screwdriver into the hole and pull the nameplate off; then loosen the screw holding the dial. If the thermostat assembly requires replacement or any leads require replacement, you will need to silver solder. To avoid deforming the thermostat contact arms from the heat of silver soldering, use a torch with a pinpoint burner to solder the connections.

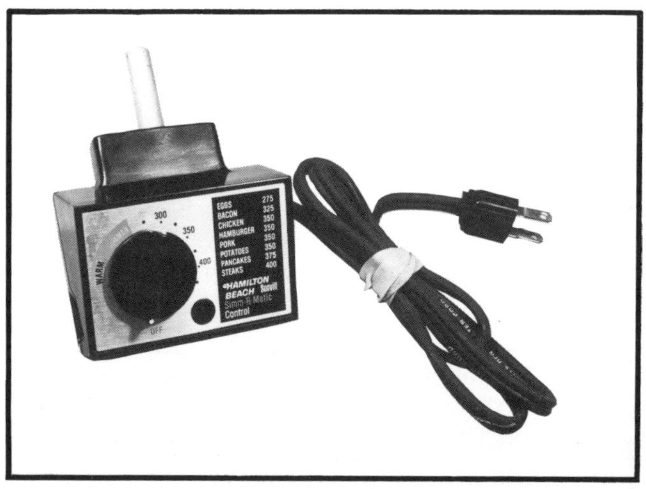

A typical thermoplug control.

To troubleshoot a thermoplug, you should have a heating element connected to it. If the neon light does not turn on, check the current-limiting resistor in series with the light by making a resistance test with the VOM. Remember, the thermoplug power cord must be disconnected from the convenience outlet when continuity tests are made. If the appliance for which the thermoplug was originally purchased is available, the degree reading on the control can be calibrated by altering the position of the knob on the shaft. The setscrew, which is often located under a piece of trim in the center of the knob, must be removed or loosened so that the knob can be adjusted. Use a thermocouple to measure the temperature in the appliance when resetting the position of the knob. Another possible problem you may encounter is that the thermoplug electrical terminals and the temperature-sensing probe often become dirty or corroded. Clean and polish the terminals and the sensing element with light sandpaper or steel wool.

If the heating element is molded into the base, as it usually is with the thermoplug type of appliance, only the element terminals are available for testing; the element is not replaceable or repairable. You can use the VOM to make a continuity test to verify any suspicion that the heating element has an open circuit. If you get a shock from the appliance, make a continuity test between each individual heating element terminal and a bare metal spot on the appliance base. If there is continuity, it indicates that a ground condition exists and the only solution is to replace the base of the appliance. Handles, legs, terminal covers, bases, covers or lids, and knobs of each of these cooking appliances are available and can be replaced.

Electric Frying Pans, Skillets, and Woks

Frying pans and skillets are similar in design and operation. They are used for frying, sautéing, and roasting. They are made of aluminum or stainless steel and are thermostatically controlled to maintain the correct cooking temperature. The heating elements normally operate at between 1,000 and 1,500 watts. Frying pans and skillets today have heating elements that are an integral part of the base. The thermostatic control is contained in the thermoplug, which is easily disconnected to allow the frying pan or skillet to be immersed in water for cleaning. On older models without thermoplugs, the thermostat is located under the pan with the bimetallic strip against the bottom of the pan; these older models are not immersible. To repair the thermostat of the older models, a metal or a plastic cover is removed to get to the internal components. The thermostat may also be located in the handle or below the handle on the frying pan. Neon indicator lights indicate when the preheat cycle is over.

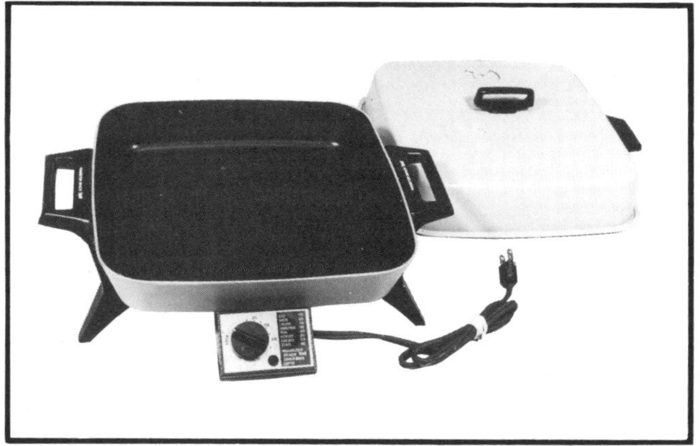

Most modern electric frying pans have thermoplugs; older models have thermostats and controls built into the handles and bases.

In modern frying pans and skillets which are equipped with a thermoplug, the heating element is nonreplaceable and nonrepairable. If the thermostat control knob becomes hard to turn, it can be lubricated with molybdenum disulfide oil. This oil can withstand the heat produced by the frying pan. If the knob will not turn at all, the frying pan, if it is an older model, or the thermoplug will have to be disassembled and checked for a mechanical breakdown. If the thermostat on the older models is replaced, make sure that new gaskets are used so that no moisture can leak into the electrical circuits.

If the skillet or frying pan does not appear to be heating properly, a thermocouple or thermometer can be used to check it. Place the thermocouple or the thermometer on the inside bottom of the pan. Keep the lid on except when making a thermometer reading. Turn the appliance on and set the heat control at about 300 degrees F. Allow the appliance to heat and to cycle through several cycles, as indicated by a neon signal light or a wattmeter or ammeter connected into the appliance circuit. With the thermostat control set at 300 degrees F, the thermostat should keep the frying pan or skillet at a temperature between 250 and 350 degrees F. If this range is exceeded, you should calibrate—if possible—or replace the thermostat control.

An electric wok is used to prepare food in the Oriental tempura style of frying. The wok is different from a frying pan in that it builds up an intense heat which allows vegetables and other finely chopped foods to be very rapidly and lightly cooked. The heating unit in the wok is located in the center area of the base to concentrate the heat there and to keep the cooking oil at the proper temperature. Because

Electric Frying Pan, Skillet, and Wok Troubleshooting Chart

Trouble	Probable Cause	Remedy
Frying pan does not heat and indicator light is off and is not cycling.	No input voltage.	Check thermostat control; then turn to maximum heat position. Recheck frying pan.
	Line cord is defective.	Check line cord; replace if necessary.
	Contacts in thermostatic control are open.	Clean and adjust points on contacts; replace thermostatic control, if defective.
Frying pan does not heat, but indicator light—if it has one—is on.	Heating element is defective.	Check heating element; replace if possible, or replace frying pan.
Frying pan has no heat control; will not shut off or overheats; indicator light does not cycle off.	Contacts on thermostatic control are shorted or stuck together.	Check thermostatic control. If contacts are shorted or fused, replace thermostat assembly.
	Thermostatic control is misadjusted.	Adjust thermostatic control.
Indicator light does not work; skillet does heat.	Indicator light bulb burned out.	Replace bulb.
	Resistor shorted out.	Replace resistor.
Frying pan blows fuse or trips circuit breaker.	Short circuit.	Check line cord, thermostatic control, and heating element for continuity. Replace shorted part or replace frying pan.
Frying pan shocks user.	Short circuit to frying pan and cover.	Visually inspect first; then check for continuity and replace shorted part or replace frying pan.

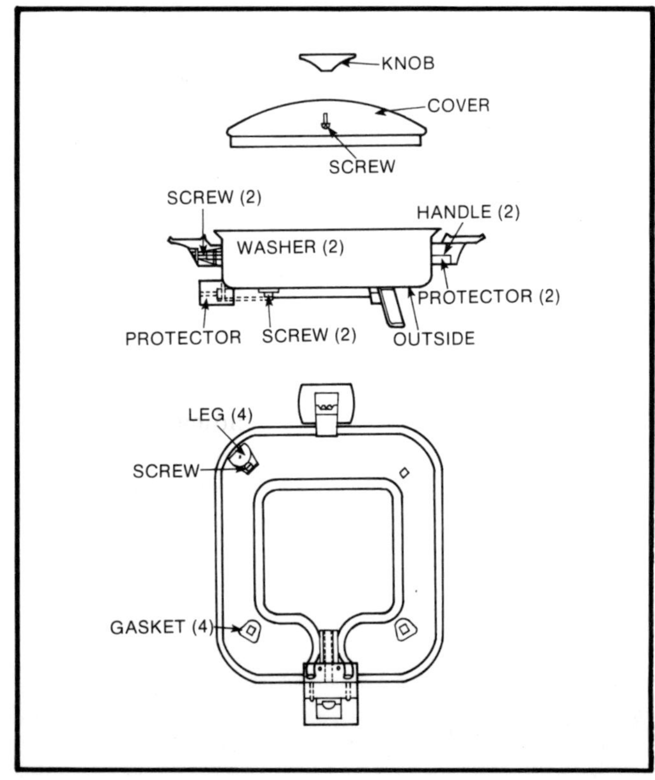

Main parts of an electric frying pan with a plug-in thermoplug.

of the intense heat at the center area, the temperature recovery of the cooking oil is almost immediate after the addition of cold foods. This allows food to be quick-fried without absorbing excessive oil. Otherwise, the wok is not electrically different from other frying pans or skillets.

Coffee Makers

There are actually two basic types of coffee makers: percolators and drip coffee makers. Both operate in a similar fashion electrically; however, they operate differently in their method of passing hot water through ground coffee beans and into a storage vessel. Basically, the percolator has a tiny well in its base which is heated, causing a small amount of water to boil and pass up through a tube, where it is deflected over ground coffee. The water seeps through the coffee and drips back into the bottom of the pot. In a drip coffee maker, a separate reservoir provides water to a heater and pump, which heats and pumps hot water to the coffee beans. The coffee then drips through into the storage vessel. All better coffee makers operate automatically. Also, once the coffee has been made, it is

The two most common types of coffee makers found in homes today are percolators and drip coffee makers.

kept heated by a warming element. We will look at these two types of coffee makers individually, because they use such different means to make coffee.

Percolators In a typical automatic coffee percolator, ground coffee beans are placed in a basket and cold water is poured into the body. The electrical power plug is connected to a convenience outlet. If provided, the strength lever used to set the coffee strength between weak and strong is set to the desired position. Small amounts of water are heated rapidly by the main heater at the base of the stem and pump assembly. The water is heated to the boiling point and a small amount of steam is produced. The steam forces the water in the stem up and out of the stem and across the perforated spreader, allowing the water to drip into the ground coffee beans. The heated water passes through the beans in the basket and through the perforated basket bottom into the lower part of the body of the percolator. When the steam forces water up the stem, there is a decrease in pressure at the stem and pump assembly. This pressure decrease allows the water, which is at atmospheric pressure, to pass through holes of the water inlet valve into the stem and pump assembly. This new charge of water is rapidly heated by the main heater until the water at the base of the pump assembly is at boiling temperature and turns to steam. Again, the steam forces the water in the stem to go up and flow over the top of the stem to the spreader. When the steam pressure builds up in the stem, it prevents additional water from coming in the water inlet valve. This cycle of admitting water into the stem and pump assembly, heating part of it to a boiling temperature, forcing the water in the stem by steam pressure up through the stem, and then admitting a new charge of water through the water inlet valve into the stem and pump assembly continues during the percolation of the coffee. The cold water that was placed into the body of the percolator eventually becomes hot as it cycles and drips back down from the basket after passing through the ground coffee beans. A thermostat senses the temperature of the water in the body. When a temperature of between 175 and 195 degrees F is reached, a thermostat bimetal strip causes the contacts to open, breaking the current through the main heater. This, in turn, causes a warming circuit to keep the coffee in the body of the percolator at a high temperature, but the temperature is not hot enough for percolation to continue. Note that it is the period of time that the cycle of percolation continues that makes the coffee have either a weak or a strong taste. If the water is too hot during the percolation cycle because the thermostat is set too high, it causes the coffee to taste bitter because of the extended percolation period. If the water is too cool, that is, if the thermostat is set too low before percolation ceases, insufficient water passes through the coffee beans and the coffee tastes weak. Besides the thermostat setting, the initial temperature of the water will greatly affect the strength of the coffee. If hot water is used—that is, poured into the coffee pot at the start—the time duration of the perking cycle is decreased and the coffee will be weak. If the coffee making process is started with very cold water, the percolation duration is increased and the coffee will be stronger.

In a typical percolator, the main heating element, which in the water-boiling element is rated between 500 and 1,500 watts, is located directly in the center surrounding the stem and pump assembly. After the initial percolation is completed, the thermostat causes the warming element to operate. The warming element is a higher-resistance element than the main heating element, and its electrical configuration in the circuit is dependent upon the manufacturer's design. When the warming element is operating, the percolator draws only about 100 watts of power, which is just enough to compensate for the heat loss caused by the natural cooling of the coffee. The warming element may run as low as 25-100 watts. Most percolators have an indicator lamp that glows when the brewing is completed. It is usually a neon lamp that is connected in series with a current-limiting resistor to prevent the neon bulb from burning out. The bottom of a percolator can be badly warped by the intense heat of the main heating element if the pot is started without any water in it or if the pot runs dry during the percolation cycling. However, if the pot has a warming element and the percolation cycling is completed when the pot runs dry, there probably will not be any damage, because of the lower temperature of the warming element.

Most coffee percolator failures can be traced to improper use: breakage occurring when the appli-

Coffee Percolator Troubleshooting Chart

Trouble	Probable Cause	Remedy
Coffee maker does not heat.	Line cord is defective.	Check line cord for continuity; replace if defective.
	Fuse is blown.	Replace fuse if blown and check for short circuit that could have caused problem.
	Main heating element is defective.	Check main heating element for continuity; replace if necessary.
Coffee maker does not percolate.	Main heating element is defective.	Check main heating element for continuity; replace if defective.
	Thermostat is open when water is cool.	Check thermostat; replace if defective.
Coffee does not stay warm.	Warming element is defective.	Check warming element; replace if defective.
Coffee maker does not stop percolating.	Thermostat is defective.	Check thermostat; replace if defective.
Coffee maker is slow to perk.	Pump jacket is loose in stem.	Tighten pump jacket.
	Stem is partially clogged with corrosion.	Clean out corrosion.
	Thermostat is defective.	Check thermostat; replace if defective.
	Main heating element is coated with lime and coffee buildup.	Clean main heating element with a vinegar and water solution.
Coffee repercolates.	Thermostat is operating erratically.	Check thermostat for continuity. Clean contacts, repair, or replace as necessary.
Coffee boils.	Thermostat is defective or out of calibration.	Check thermostat; if not defective, recalibrate thermostat to operate at 175 to 190 degrees F.
Coffee is not hot enough and is weak.	Thermostat is defective.	Check thermostat; replace if defective.
	Pump jacket is loose on stem.	Tighten pump jacket.
	Pump does not seat properly.	Reseat pump.
Coffee is too strong.	Taste control is incorrectly set.	Reset taste control.
	Residue is inside pot.	Clean inside of pot.
Coffee is very weak.	Did not use enough ground coffee.	Add ground coffee.
	Started making coffee with hot water.	Always use cold water instead of hot water.
	Taste control is incorrectly set.	Reset taste control.
	Pump valve is stuck.	Clean pump valve, or replace if necessary.
Coffee maker blows fuse or trips circuit breaker.	Short circuit.	Check line cord for continuity. Make a visual inspection of the heating circuits. Make continuity tests until short circuit is found; replace defective parts if possible.
Coffee maker shocks user.	Short circuit to outside of pot.	Visually inspect. Check for continuity until short circuit is found; replace defective parts if possible.

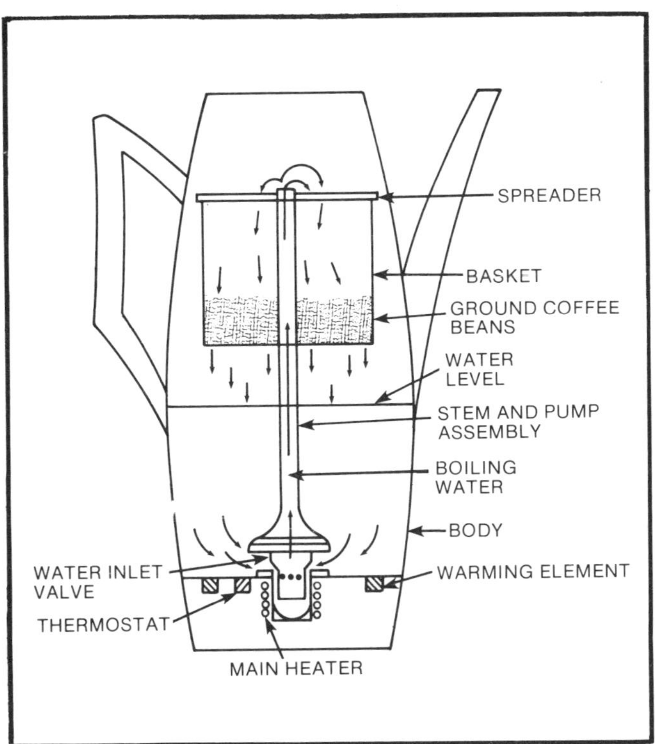

Cross section of a typical coffee percolator.

ance is dropped, dry operation, and water damage. Overheating can cause damage beyond practical repair, especially if the metal is so badly warped as to make sealing between the heating element and the pot impossible. Repairs are easily made if damage is confined to the heating element or to other easily replaceable electrical parts, or if a fuse has protected the appliance. Generally, coffee percolator pots cannot be immersed in water for cleaning. A warning to that effect is on the nameplate. Those coffee pots which can be immersed are equipped with special seals. If a coffee pot shows signs of rust, the rust-damaged parts should be replaced. On models designed to be immersed, evidence of rust or water inside the control enclosure should lead to a thorough inspection of seals and replacement as necessary. Aside from obvious mechanical damage, a coffee pot is faulty if it will not heat up, if it heats up improperly, or if it cannot be adjusted by the taste control lever. These faults, all electrical, are easily isolated to the defective component by resistance measurements. Be sure the appliance is unplugged before removing the bottom cover to gain access to the electrical parts.

Drip Coffee Makers In the past few years, the automatic drip coffee maker has become extremely popular for home use. This type of coffee maker is rather different from the percolator, with the greatest practical difference being that it can make coffee very quickly. For example, an average small drip coffee maker can make up to eight cups of coffee in less than seven minutes. To operate a drip coffee maker, cold water is poured into a reservoir, and the main power switch is turned on. Do not turn the automatic drip coffee maker on unless there is water in the reservoir. From the reservoir, the water is fed by gravity to an internal tank heater, where it is rapidly heated and pumped through a tube to a dripper spout. A conical basket distributes the hot water around the ground coffee. The hot water picks up the flavor of the coffee beans as it passes through a disposable filter and then into a glass storage carafe. The warming plate under the carafe keeps the coffee at proper serving temperature.

The electrical circuit of the drip coffee maker is similar to other coffee makers. When the main power switch is turned on, AC current flows through both the warming and the main heating elements, which are connected in parallel; current also lights the indicator lamp to show the user that the coffee maker is on. When the brewing cycle is completed, the thermostat contacts open, breaking the current path to the main heating unit. The warming element located in the warming plate of the coffee maker keeps the coffee warm. On some models, there are separate ON-OFF switches for the main heating element and the warming plate. On these models, the main heating element must be turned off manually when the brewing cycle is finished.

Electrically, drip coffee makers do not add any new problems or configurations. The circuits and individual components can be checked by voltage or continuity tests if the coffee maker is not working. However, one problem which is prevalent with this type of coffee maker is mineral buildup on the brewing components. Water used for making coffee contains, to varying degrees, dissolved minerals that will affect the operation of the drip coffee maker if not taken care of. When heated, the minerals deposit on the main unit, in the tubes, and in the spout. The rate at which the deposit forms varies according to the hardness of the water used and the frequency with which the coffee maker is used. The efficiency of the coffee maker will eventually be impaired, excessive steaming will occur, and unpumped water will remain in the reservoir if the minerals are not removed. Special cleaning with vinegar is recommended about once a month.

To clean the drip coffee maker, pour household vinegar into the reservoir until it is half full. Then, add cold water to the vinegar until the reservoir is filled. Replace the reservoir cover. Place the basket in its proper position and put the carafe on the warming plate. Place the power switch to ON. Pump about three-fourths of the solution into the carafe. Then turn the power switch to OFF. Return the hot solution from the carafe to the reservoir, adding it to the cold solution in the reservoir. Replace the reservoir cover and let the solution stand for 15 minutes. To recycle the solution, it is necessary to leave some

Drip Coffee Maker Troubleshooting Chart

Trouble	Probable Cause	Remedy
Coffee maker does not pump.	Pump unit is burned out.	In future, make sure there is water in the reservoir before you turn on the coffee maker.
	Defective part or parts; circuit is open.	Check line cord, switch, thermostat, and main heating element for continuity. Replace defective part.
	Hose is pinched.	Repair or replace hose.
Coffee maker does not pump all water out of reservoir.	Hose is damaged.	Check hose for pinches. Repair or replace.
	Coffee-making unit or spout is dirty.	Clean using vinegar solution.
	Thermostat is defective.	Check thermostat for continuity; replace if defective.
	Main heating unit surges.	Replace main heating unit.
Coffee maker leaks water down the front.	Coffee-making unit and spout out of adjustment.	Readjust unit and spout. If water still leaks, replace spout; if water leaks at this point, replace unit.
Coffee maker does not restart on second brew cycle.	Thermostat is defective.	Check thermostat; replace if necessary.
	Hose is pinched.	Repair or replace hose.
	Warming bracket assembly loose.	Tighten warming bracket assembly.
Coffee maker restarts while warming.	Warming unit is defective.	Check warming unit; if it shows no continuity or if the plate does not get warm, replace it.
	Warming bracket assembly is loose.	Make sure warming bracket is tight against heating plate and bracket arm is tight against insulation and unit.
Warming plate does not warm.	Warming plate heating element is defective.	Check warming plate element for continuity; replace if defective.
Coffee maker is pumping lime.	Coffee-making unit is dirty.	Clean using vinegar solution.
Coffee does not stay hot.	Warming bracket assembly is loose.	Check assembly of the warming bracket and make sure it is tight so that it conducts heat to the plate.
Coffee too weak at a low cup requirement.	Ground coffee not spread evenly in filter.	Spread ground coffee evenly.
Coffee not hot enough at a low cup requirement.	Glass carafe is too cold.	Let carafe and brewed coffee sit for a short time on warming plate, allowing coffee to warm, because a cold carafe—just rinsed out with cold water, for example—takes too much heat out of the coffee.
Coffee maker blows fuse or trips circuit breaker.	Short circuit.	Check line cord for continuity. Make visual inspection of heating elements. Make continuity tests until short circuit is found; replace defective parts.
Coffee maker shocks user.	Short circuit to body of coffee maker.	Inspect visually. Check for continuity until short circuit is found; replace defective parts.

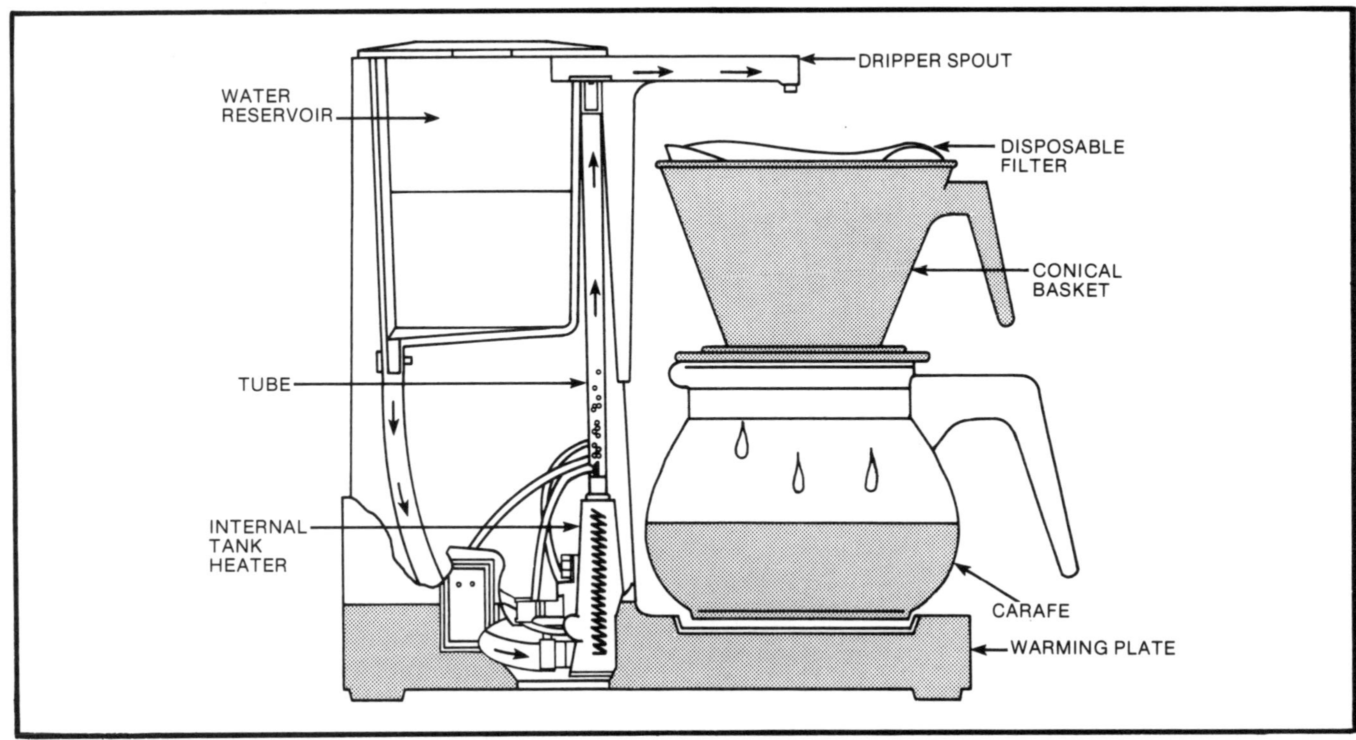

Cross section of a typical automatic drip coffee maker.

of the solution in the reservoir each time. Pump only about three-fourths of it into the carafe until the final cycle, at which time all of the solution should be pumped into the carafe before the power switch is turned off. You can save the vinegar-water solution; it may be used for three or four cleanings.

When you have pumped all of the vinegar solution out of the coffee maker, unplug the cord from the convenience outlet and detach the spout. Rinse the reservoir with tap water and invert the coffee maker to remove all of the water. Also be absolutely sure that all of the vinegar solution has been cleaned out of the unit. For a final rinsing, replace the spout and fill the reservoir halfway with clear, cold water. Replace the reservoir cover and plug the power cord in. Place the power switch to ON to pump the water through the basket-spreader assembly into the carafe. Place the power switch to OFF and the cleaning job is done. Normally, one cleaning cycle will be sufficient. It may be necessary to run the vinegar solution through the coffee maker two or three times if you have noticed excessive steaming and unpumped water remaining in the reservoir when you make coffee, or if the coffee maker has not been cleaned for a long time.

Automatic Toasters

Automatic toasters are difficult small appliances to repair; electrically, they are not much different from other small appliances and are not difficult to repair in that respect. But mechanically, there are many different kinds of parts and a wide range of systems for popping the toast up when it is done. There are also many manufacturers and many models to further complicate matters. Some of the parts that may be within a toaster are the bread carriage, the main power switch, outside and center heating elements, thermostat, bimetal strip, bimetal heater, carriage release lever, dashpot or flywheel assembly, toast color-setting lever, and bread crumb tray.

You are already familiar with the operation of a toaster just from operating your own. With automatic toasters, bread is placed into the bread carriage tray. Then, a bread carriage lever is depressed, which sends the bread and tray to the bottom of the toaster, where the tray is latched by a latching mechanism. The bread is positioned in parallel with the heating elements. When the bread carriage lever is mechanically latched into position, a main power switch consisting of two sections, one in each leg of the AC power input, closes and applies 120 V AC to the heating elements—a center element and two outside elements for a two-slice toaster.

When the bread carriage lever reaches its bottom position, not only the heating elements turn on through a power switch, but a timing mechanism also starts to operate. After a certain amount of time elapses, which is made variable by changing the setting of the color-setting lever, the timer causes the carriage release lever to release the bread carriage. A spring mechanism then raises the bread

An automatic pop-up type of toaster.

carriage and bread up to the top of the toaster. To prevent the toast from literally flying out of the toaster, the bread carriage is controlled by a dashpot or flywheel assembly, which allows the carriage to rise slowly. When the bread carriage is released, the heating elements are turned off by the main power switch. Toasters operate at approximately 1,000 to 1,100 watts.

The basic toaster operation includes five different mechanisms: power switch, heating elements, timing mechanism, latching and unlatching mechanism, and a controlled mechanical pop-up mechanism. The power switch has two sections that apply the input power from both sides of the convenience outlet to the toaster circuit. The heating elements may consist of a parallel or a series-parallel circuit. The center element normally is connected by itself; the outside elements are connected either in series or in parallel. The power rating for the outside heating elements is greater than that of the inside element; the reason for this is that the reflected heat on the inside element effectively increases its heat by about 10 percent. Therefore, to toast the bread evenly on both sides, the amount of heat required on the center element is less than that required on either of the outer elements.

The timing mechanism determines the color of the toast. This color is determined by the amount of time the toast is left in the well with the heating elements on. Bimetal strips are used in toaster designs, but they are not used as thermostats to control the temperature. The current through the heating elements of a toaster remains the same and toaster elements remain on from the initial time they are turned on to the end of the toasting cycle. Bimetallic strips are used to vary the amount of time that the toast remains in the carriage in the lowered position with the heating element on. The bimetallic element bends, causing a mechanical system to unlatch the bread carriage, while the release system turns the main power switch to OFF, breaking the path of current to the heating elements.

The latching and unlatching mechanisms are mechanical and in some cases electromechanical. When the bread carriage lever is depressed, it is mechanically engaged by the carriage release lever. This holds the bread carriage down until the timing mechanism causes the carriage release lever to unlatch the latching mechanism, allowing springs to pull the bread carriage upward in the toaster. In some designs, a timing mechanism causes an electrical solenoid or relay to energize, which pulls a lever to unlatch the latching mechanism. Another type of latching and unlatching mechanism works by means of a magnet which works in conjunction with the bimetal heating element of the thermostat timing device.

Toasters contain pop-up mechanisms that control the speed of the rise of the bread carriage to prevent the toast from flying out of the carriage. These assemblies vary on different models and, as mentioned before, may be flywheel assemblies, dashpots, or pneumatic systems on some older models.

When repairing a toaster, sometimes the easiest approach, if the manufacturer's directions are not available, is to remove the end of the toaster where the bread carriage and the controls are housed. By removing the end, you can observe the operation of the toaster and try to visualize any problems. The

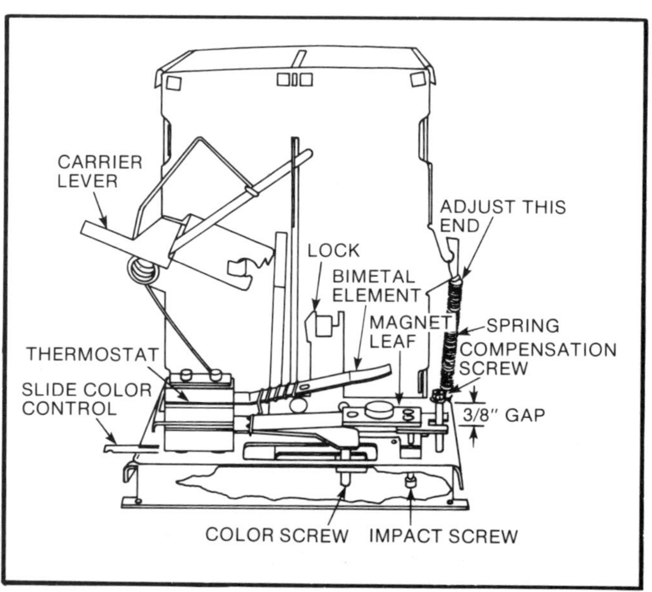

Cross section of a typical automatic pop-up toaster showing mechanical latching-unlatching system.

Automatic Toaster Troubleshooting Chart

Trouble	Probable Cause	Remedy
Toaster does not heat.	No input voltage.	Depress bread carriage lever and ensure that it latches and remains latched.
	Line cord is defective.	Check line cord for continuity; replace if defective.
	Main power switch is defective.	Check both sections of main power switch for continuity. Contacts must be clean and touch with adequate pressure. Replace switch if defective.
	Loose connections.	Check for any loose connections. Clean and tighten.
	Switch contacts do not stay closed when the operating lever is depressed.	Check the latch and lifter arm assembly. Adjust the lever to make good contact or replace the assembly.
One or more heating elements do not work.	Heating element is defective.	Visually inspect to determine if a heating element is defective. If visual inspection does not show a defective element, check for continuity. Be sure to watch for parallel elements that may give a false indication of continuity.
Heating elements do not go off.	Main power switch is defective.	Check main power switch; replace if defective.
	Carriage latch mechanism is defective.	Inspect mechanism. Repair or replace.
	Bimetal strip is bent or broken.	Check bimetal strip. Repair or replace.
Toast does not pop up when done on toasters with bimetal heater.	Bimetal thermostat is defective.	Check bimetal thermostat for continuity.
	Bimetal heating element is defective.	Check bimetal heating element for continuity.
Toast does not pop up when done or toast is not the color selected on toasters without bimetal heater.	Timing mechanism is defective.	Check bimetal strip for bends or breaks; repair or replace.
		Check electrical solenoid.
		Check timer spring.
Toast flies out of toaster.	Dashpot is defective.	Tamp around the edge of asbestos packing to close air leaks.
	Flywheel assembly is defective.	Inspect flywheel for bent or broken parts; repair or replace.
Toast rises too slowly.	Dashpot is defective.	Make a pinhole in the asbestos packing.
	Flywheel assembly is defective.	Inspect flywheel for bent or broken parts; repair or replace.
	Springs are too weak.	Replace springs.
Bread carriage does not latch down.	Mechanical latch is broken or bent.	Visually inspect; repair or replace.

Automatic Toaster Troubleshooting Chart (Continued)

Trouble	Probable Cause	Remedy
Toast is not toasted to color selected.	Bimetal strip is bent or broken.	Check bimetal strip; repair or replace.
	Bimetal strip and magnet misadjusted.	Adjust setscrew.
One side of toast is untoasted.	Heating element is defective.	Replace heating element on side where toast is not being toasted.
Toast burns.	Switch is not working properly.	Clean; if defective, replace.
	Timer or thermostat calibration improperly set.	Adjust or replace parts if defective.
	Bimetal strip distorted.	Check; replace if strip cannot be straightened.
Crumbs, butter, raisins, and other debris accumulated in toaster.		Clean bread tray in soapy water. Remove charred particles on heating elements with a small paint brush. Remove grease with solvent.
Toaster blows fuse or trips circuit breaker.	Short circuit.	First test line cord for continuity. Check heating elements visually; then make continuity tests until short circuit is found.
Toaster shocks user.	Short circuit to toaster case.	Inspect visually for short circuit; then make continuity tests until short circuit is found.

electrical circuit of a toaster can be checked out by means of continuity tests or voltage tests similar to those used for other appliances in this book. You can usually determine that one of the heating elements is not working correctly by a visual inspection. If an element is not operating correctly, it must be replaced. Dashpot adjustments are easily made. If the bread carriage rises too fast, tamp around the edge of the asbestos packing on the dashpot. This tightens the fit in the dashpot. If the carriage rises too slowly, make a pinhole in the asbestos packing.

Calibrating a toaster means to set the timing mechanism so that a certain toast color results when the color selector is set. For example, when the color selector lever on the toaster is set to a medium brown, the toast should appear as a medium brown color to match the color band shown on the toaster—if color bands are provided. If the color does not match, the toaster can be calibrated by the adjustment of a setscrew to compensate for the inaccurate time. You should refer to the manufacturer's calibration procedure for making this timing setting. Without a manufacturer's calibration procedure, you can adjust the setscrew of the toaster being calibrated by rotating it about one-half turn and then trying a piece of toast and matching its color with the color selector. In some toasters the setscrew rotates clockwise to increase the speed of toasting; in others counterclockwise rotation increases the speed. This is something you will have to determine for yourself.

Toasters should only be lubricated as directed by the manufacturer. Some manufacturers require lubrication on all moving parts; care should be taken, however, to prevent the lubricant from running down into the bimetal assembly. Lubricate with a graphite grease stick and lubricate contact surfaces such as on the latch lever and the spring blade of the contact points. A brush-on type of graphite lubricant can be used on the threads of the main control shaft, the carrier lever, and the bread carrier slide.

One of the main causes of trouble in a toaster is the accumulation of crumbs, butter, raisins, and similar particles of food. A good form of preventive maintenance is to periodically clean these particles from the toaster. You should first remove the bread tray, which can be cleaned in soapy water. Crumbs that become attached to the heating elements become charred and can be brushed off by using a small, stiff paintbrush. Grease can be removed by applying a grease solvent to a rag and gently wiping the affected areas. Inspect the magnet surface. Remove any foreign particles on the magnet that could cause erratic operation. If arcing was noticed between the bimetal strip and the magnet, replace the bimetal strip. If there is uneven contact between the bimetal strip and the magnet, you can sometimes twist the bimetal strip to readjust it. If the strength of the magnet has deteriorated because it is old or if it has been extremely overheated, replacement is the only remedy.

Hair Setters

A hair setter consists of a number of different-sized plastic hair curlers that have aluminum interiors for retaining heat. The curlers fit on heating posts during a warmup period. Electrical current flows through a heater which heats a metal plate. The metal plate then heats the heating posts, which, in turn, heat the aluminum interiors of the curlers. The curlers retain this heat which is then utilized to curl damp hair wound around the curlers. Curlers normally have an indicating dot on the top that is usually red when the curlers are cool and turns black when the curlers are ready for use. Some hair setter units also utilize a steam system. A heating element creates steam which, in turn, heats the curlers. No aluminum interiors are used in these models.

Hair setters can have one of several circuit configurations. For example, in one, electrical power is applied as soon as the plug is connected to a convenience outlet. The 120 V AC current travels through the thermostat and a fuse to the heating element, which heats the curlers as previously described. Since these components are wired in series, a malfunction in the thermostat, the fuse, or the heating element causes the complete circuit to stop operating. The thermostat in this type of hair setter maintains the temperature of the heating posts at a constant level.

In a slightly different arrangement a switch, which is a microswitch, has normally open contacts that close when the lid of the hair setter is opened. Therefore, when the user opens the lid, the microswitch closes and electrical current is applied to the circuit. Current immediately flows through the microswitch and through a resistor and white light, which indicates that power is being applied to the hair setter. Since the hair setter is at room temperature, the thermostat contacts are normally closed and an electrical current also passes through the contacts to the parallel circuit, consisting of one leg that has a resistor and red light and a second leg that has a fuse and the heating element that heats the posts of the hair setter. The red light lights when the thermostat contacts are closed; therefore, the user can look at the red light and know that the hair setter has reached its operating temperature when the red light goes off. From that point on, the thermostat turns the current on and off to maintain a constant temperature at the curler heating posts. The fuse is a protective device that opens the heating element circuit in the event that the thermostat contacts become welded closed, causing an excessive amount of heat to be produced within the hair setter. This type of hair setter operates at about 400 watts of power and is ready for use after about four minutes.

In a third common configuration, the circuit is energized as soon as the plug is connected to a convenience outlet. In this configuration, current flows through a series circuit consisting of a fuse, two heaters, and a thermostat having normally closed contacts when the temperature is below the operating temperature of the hair setter. A neon light is in parallel with the thermostat, and when the thermostat contacts are closed, the neon light is short-circuited out of the circuit. Therefore, the light comes on when the curlers have reached the correct temperature. When the curlers are at this temperature and the thermostat contacts are open, there is a series circuit through the fuse, one heater, the neon light, and the second heater. Since the neon light drops approximately 60 volts across it, only 60 volts remain to be divided between the two heaters, so the current flow is decreased and the heaters operate at reduced power to maintain the temperature within the hair setter.

Troubleshooting a malfunctioning hair setter consists of the usual continuity and voltage tests. In the second circuit configuration mentioned above, the indicator lights really do the troubleshooting for you. For example, if the hair setter is turning on and the red light does not come on but the white light does, you can immediately determine that there is electrical power to one side of the thermostat but no power on the other side. Therefore, you can conclude that the thermostat contacts are open. In a similar manner, if the red light continually operates, there

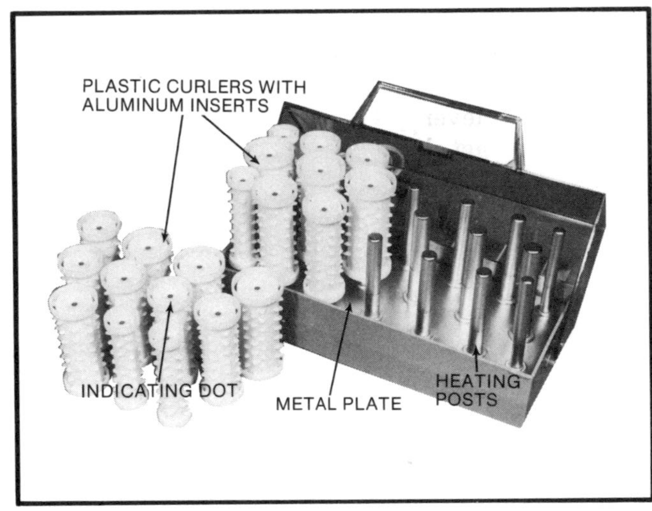

A typical home version hair setter.

Hair Setter Troubleshooting Chart

Trouble	Probable Cause	Remedy
Hair setter produces no heat.	No input voltage.	Place ON-OFF switch to ON position or open lid to close microswitch.
	Fuse is open or circuit breaker is tripped.	Check fuse or circuit breaker; replace or reset if blown. Determine what caused the overload.
	Line cord is defective.	Check line cord for continuity; replace if defective.
	Thermostat is open.	Check thermostat for continuity. Clean contacts; replace if still not working.
	Heating element is defective.	Inspect heating element; replace if defective.
Hair setter produces heat but indicator light does not work.	Light is defective.	Check indicator light; replace if necessary.
Heat circuit does not cycle on and off.	Thermostat is short-circuited.	Check thermostat for continuity; replace if VOM reading is zero or very close to zero, indicating a short circuit under these conditions.
Hair setter blows fuse or trips circuit breaker.	Heating element is short-circuited.	Check heating element for continuity; replace heating element if short circuit condition exists.
	Internal short circuit.	Make continuity tests until short circuit is found.
Hair curlers are dirty.		Clean hair curler with a soft brush and a mild solution of detergent and water.

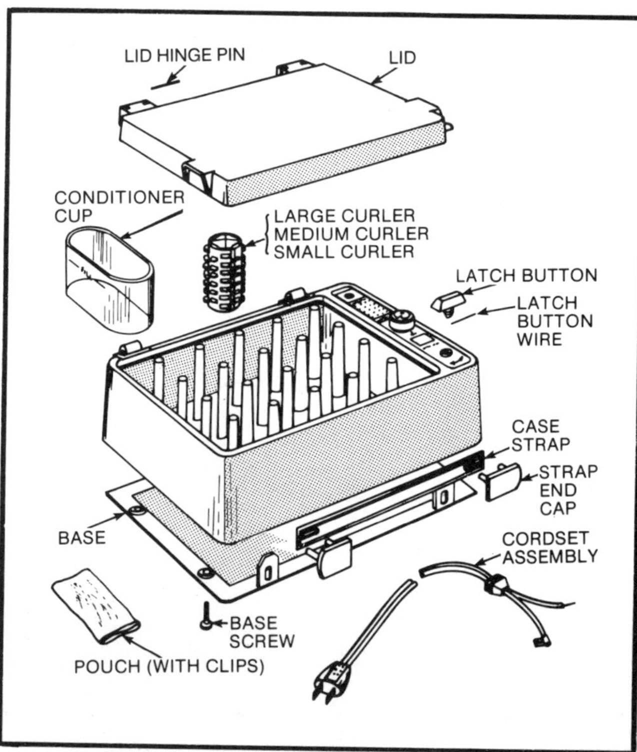

View of a typical disassembled mist-type hair setter.

is no cycling of the thermostat. This points out a condition in which the thermostat contacts are permanently closed or short-circuited. And finally, if the red light cycles on and off but there is no heat in the hair setter, you can determine that either the fuse or the heating element is open. A continuity check can be made of the fuse and the heating element after the power plug has been removed from the convenience outlet. If the hair setter is experiencing problems, you can also check that it is operating at the proper power rating and that the operating temperature is correct.

Hair curlers should be cleaned with a soft brush and a mild solution of detergent and water. Drain the curlers dry by placing them on a paper towel with the signal dots upward. Wipe a hair setter clean with a clean cloth dampened with the same mild solution. Do not immerse a hair setter in water.

Hair Dryers

Hair dryers are primarily used to blow hair dry with heated air, although ingenious homeowners

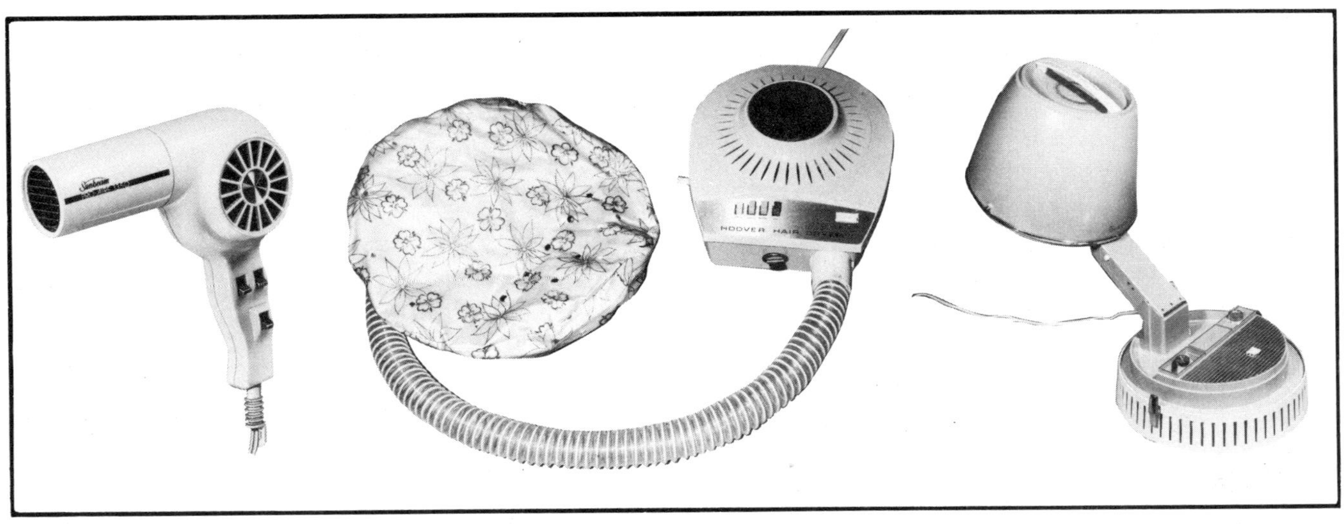

There are several types of hair dryers: hand-held models (left); models with a bonnet that fits over the head (center); and models that use a plastic hood (right).

seem to have found a whole slew of other ways to utilize the dryer's stream of warm air around the home. There are basically three types of hair dryers: hand-held dryers, dryers that sit on a table and have a hose that connects to a bonnet placed around the hair, and dryers that sit on a table and have a hard-framed plastic hood that is placed over the hair. The model with the hood is primarily a professional hair dryer, but smaller models are also made for home use. Each of the hair dryers operates in basically the same way, and none are especially complex.

Strangely enough, hair dryers are similar in operation and electrical circuitry to space heaters. In the space heater, the heating elements produce heat and a fan circulates the air into the surrounding area. In the hair dryer, the heat is also generated by heating elements, and a fan or impeller causes the air to flow directly to the hair or through a hose to a hood or bonnet and then to the hair. A dryer consists of a safety thermostat, a fan motor which is usually a shaded-pole motor that turns a metal or plastic blade or impeller, heating elements, and switches that control the on and off condition of the fan motor and the heating elements. Hair dryers do not have thermostats for controlling temperatures, only the safety thermostat. The heating elements produce heat and the motor and fan assembly blows the air through the elements; the elements continually operate at maximum heat.

When the dryer is plugged into a convenience outlet, power is coupled through the safety thermostat's normally closed contacts to the motor switch. The motor switch may be an individual switch, but it is more likely part of an interlatching pushbutton function switch or a rotary switch designed so that whenever the heating element is on, the motor must also be on. When the motor switch is closed, AC power flows through the safety thermostat and the motor switch to the motor, which drives the fan and blows cool air out of the fan. When low heat is desired, the Low switch is closed. This switch couples AC power through a low-temperature 100-watt heating element; the circuit is completed back to the other side of the AC line. If medium heat is desired, the Medium switch is closed; this action automatically opens the Low switch so that only the Medium switch is closed and only the medium heating element is on. Current now flows in a parallel circuit through the motor and the 200-watt medium-temperature heating element. When you want to get the highest heat from the dryer, the High switch is closed. This switch actually has two sections that close and connect both the 100- and 200-watt heaters in parallel with the motor circuit. The heating power used, therefore, is 300 watts and the total power drawn from the convenience outlet is 300 watts, plus approximately 30 watts for the motor.

If the hair dryer fan should fail, causing high internal heat, or if the inlet or outlet air ports of the hair dryer are blocked so that heat builds up inside the hair dryer, the safety thermostat contacts open, breaking the path of current to the heating elements and to the motor switch. After a cooling-off period of half a minute or so, the safety thermostat contacts again close, permitting current to flow to the motor and to the switch circuits for the heating elements. In some hair dryers, the safety thermostat is located between the switch wiring on the one hand and the motor switch and motor wiring on the other. With the safety thermostat in this location, the fan motor continues to operate after the safety thermostat contacts have opened to interrupt electrical power to the elements; this allows the fan to aid in cooling the overheated condition.

There are several types of malfunctions and symp-

Hair Dryer Troubleshooting Chart

Trouble	Probable Cause	Remedy
Heaters and fan motor do not work.	No input voltage.	Turn power switch on.
	Line cord is defective.	Check line cord for voltage or continuity; replace if defective.
	Safety thermostat is open.	Check thermostat for voltage or continuity. Clean contacts; if heaters and fan motor still do not work, replace safety thermostat.
	Motor switch is defective.	Check motor switch for voltage or continuity.
Heaters work, but fan motor does not.	Motor switch is defective.	Check motor switch for voltage or continuity; replace switch if defective.
	Motor coil is open.	Check coil for continuity; replace motor if coil is bad.
Heating elements do not work.	Switch or switch section is defective.	Check entire switch for voltage or continuity; clean contacts. Replace switch if defective.
	Heating element is defective.	Check heating elements for continuity; replace if defective.
Hair dryer does not shut off.	Motor switch is short-circuited.	Check motor switch for continuity; replace if VOM reading is zero or very close to zero, indicating a short circuit under these conditions.
Safety thermostat does not open when airflow is blocked.	Safety thermostat is defective.	Check safety thermostat; replace if defective.
Hair dryer blows fuse or trips circuit breaker.	Short circuit.	Conduct continuity tests until short circuit is found; replace defective parts.
Hair dryer shocks user.	Short circuit to hair dryer case.	Make visual inspection first for short circuit. Then make continuity tests until short circuit is found; replace defective part. Otherwise, replace hair dryer.

toms that are fairly common in hair dryers: the fan motor and heaters not operating, the fan motor not operating, the heaters not operating, or just one of the heaters not operating. By using voltage tests, visual inspections, and continuity tests with the power plug removed from the convenience outlet, you can troubleshoot to locate the malfunctioning component. For example, if the heating elements and the motor are both inoperable, you might suspect that either the motor switch is not operating, that the safety thermostat contacts are open, or that there is a break in the line cord. You can check these with the VOM by first checking the line cord for voltage. An indication of 120 V AC indicates that the line cord is good. The VOM is then used to verify that the safety thermostat contacts are closed as they should be under normal conditions; if 120 V AC is not present, the safety thermostat contacts are probably open. On some hair dryers the safety thermostats must be manually reset; be sure to determine whether or not your safety thermostat on your hair dryer should be reset manually. You can check the contacts by removing the plug from the convenience outlet and connecting a jumper wire between the contact connecting lugs to short-circuit and thus bypass the contacts. When the plug is reapplied, 120 V AC should be present; if it is, you have verified that the safety thermostat contacts are not closing. Remove the power plug from the outlet and inspect the thermostat. You may be able to clean the contacts using an aerosol-spray cleaner. If this does not remedy the problem, the safety thermostat must be

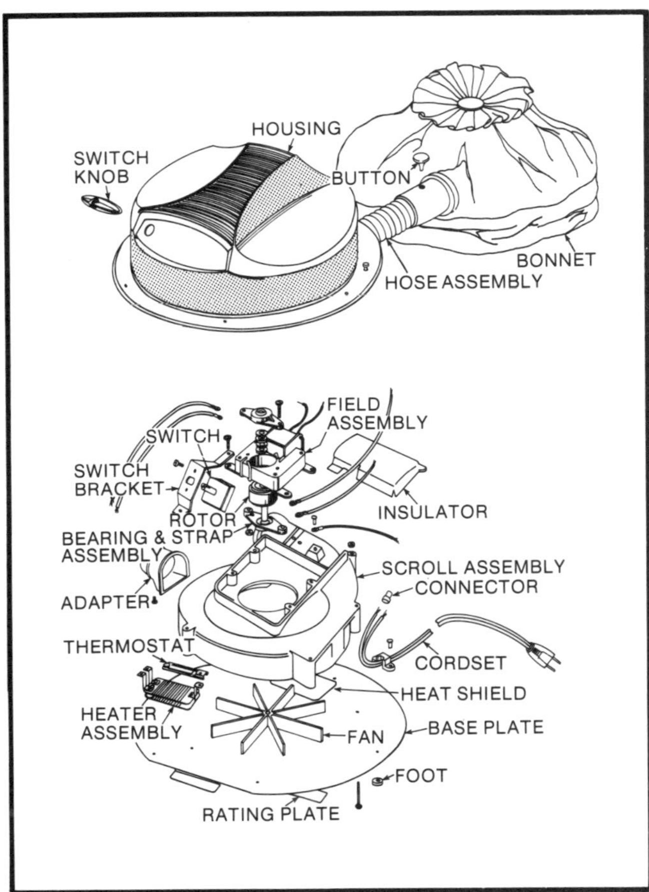

Exploded view of a typical bonnet-type portable hair dryer.

replaced. The next check to make with the VOM is the motor switch. With the power plug connected, the indication should be 120 V AC if the motor switch is closed. If 120 V AC is absent, depress the switch and see if 120 V AC is now present. Continued absence of voltage indicates a faulty motor switch that must be replaced.

Suppose the heater fan is operating but neither heating element is. Since the probability of a malfunction occurring in both heating elements simultaneously is rare, the problem is probably a broken connection at the switch assembly. Make a visual inspection first and then check for 120 V AC at the various switch connections. If the voltage is not present, there is a break in the wiring.

After a hair dryer has been repaired, a final test should be made. Check the power with an ammeter, wattmeter, or VOM, if it has this capability; the power should be within ± 10 percent of the nameplate power. With the hair dryer assembled, you can check the safety thermostat operation by blocking off the air outlet. With the heating element on, the safety thermostat should open, causing the heating element and perhaps the fan, depending upon the electrical wiring of the fan, to go off, indicating that the safety thermostat did in fact open.

Clothing Irons

There are three types of hand clothing irons used to iron fabrics: nonautomatic, automatic, and the steam iron. A nonautomatic iron consists simply of a heating element in the soleplate or base of the iron. A line cord that is internally connected to terminals within the iron is connected to the house convenience outlet. When the line cord plug is in the convenience outlet, the iron is on; when the plug is removed, the iron is off. Electrically, the automatic iron and the steam iron are identical and consist of a thermostat with normally closed contacts in series with a heating element. When a temperature lever is advanced, it closes the thermostat contacts, applying power to the iron. Electrical power from the convenience outlet flows through the closed contacts of the thermostat to the heating element, which produces the heat that warms the soleplate of the iron by conduction. The thermostat senses the temperature of the soleplate. When the proper ironing temperature as selected on the temperature lever is reached, the thermostat contacts open, breaking the path of electrical current to the heating element. As the iron cools, the thermostat closes again, and the iron again heats up to the selected ironing temperature. The variance in temperature between the closing and the opening of the contacts is about 5 degrees F. The temperature range of the soleplate varies from about 245 degrees F for rayon to 520 degrees F for linen. The steam temperature is approximately 350 degrees F.

The most common iron today is the automatic

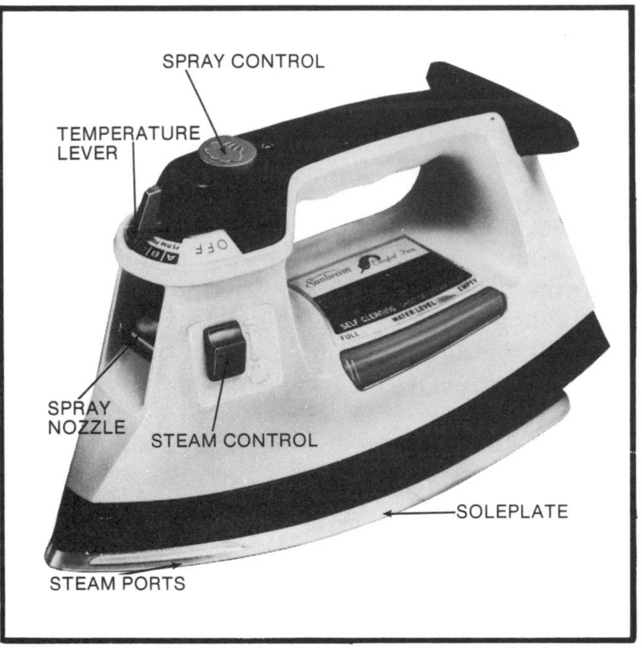

A modern automatic steam iron has thermostatic temperature control, a steam-use option, and the ability to spray.

steam iron. A thermostat automatically controls the temperature as selected by the temperature lever. The temperature lever normally has indications for rayon, silk, wool, cotton, linen, and for the steam position. The iron allows steam to come out of holes or ports in the soleplate to aid in pressing fabrics. In addition, many models contain a spray control that can be depressed to cause a spray of water to come out from a pressure nozzle in the front of the iron onto the fabric. These features aid in making the iron more adaptable to various types of fabrics.

The main parts of an automatic steam iron are the heating element, soleplate, heel plate, handle, hood, thermostat, pressure plate, jet steam control, spray control, and electrical cord. The heating elements may be nichrome ribbon, nichrome coils, or Calrod. The nichrome ribbon is wound on mica sheets and is replaceable; it is placed on top of the soleplate. This type of heating element, however, is found in older model irons. The element is held in place by the pressure plate, which keeps the heating element against the soleplate. If an insulator is required, an asbestos liner is placed between the soleplate and the pressure plate.

The nichrome coils or the Calrod are usually built into the soleplate; if a malfunction occurs in the heating element, the soleplate must be replaced. Sometimes round nichrome wire is formed as a small, round or rectangular rod and is placed in a channel in the soleplate. This type of element can be replaced, but it is recommended that the soleplate be replaced as well. The heating elements are often broken—in other words, open-circuited—because of the stress of the heavy current flowing through the element coils each time the iron is turned on and because of the continual ON-OFF cycling of the current through the element as required by the thermostat. The coils of the elements are sometimes short-circuited together or the element is grounded to the case of the iron when the iron is dropped on the floor. The electrical connections at the heating element sometimes become weakened and oxidized. The oxidation may cause the eventual breaking of the connection or the connection may become so thin that a lot of heat is produced at the connection rather than at the heating element. In all cases, it becomes necessary to replace the electrical connecting wires.

Most soleplates built today are of aluminum, which is a rather soft metal and, therefore, scratches and dents easily. For the most effective ironing, the soleplate bottoms should be flat, free of scratches and dents, and free of dirt. Sometimes the soleplate is discolored by excessive use of starch. Scratches, dirt, and discoloration from starch can be removed by buffing and polishing. If there are large dents in the soleplate, the plate should be replaced. The heel plate is at the back of the soleplate and provides a resting point for the iron when the iron is tilted back so that it rests on the heel plate and the rear of the handle.

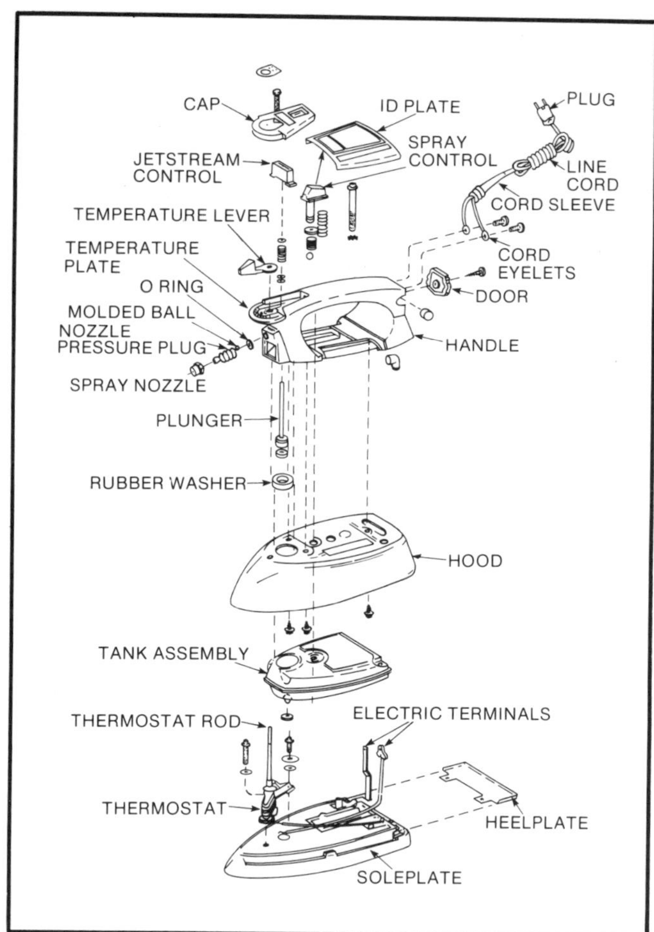

Exploded view of an automatic steam iron.

Handles manufactured today are generally plastic. The handle, which mounts on the hood, is used by the operator to move the iron back and forth across the fabric and also houses the temperature control dial and lever, the steam control, and the spray control. A stainless steel hood is located beneath the handle to shield the operator's hand from the heat of the iron and also to reflect the heat down toward the soleplate to aid in minimizing heat losses.

The thermostat in an iron consists of two arms that contain silver electrical switch contacts. These arms are termed the upper arm and the lower arm. The upper arm is movable by the adjustment of a temperature-setting control shaft and a calibration adjustment. The temperature-setting control shaft is rotated by the temperature lever on the top of the iron. The calibration adjustment setscrew is set at the factory or should be reset when a thermostat is replaced. Another arm that is parallel to the upper arm and the lower arm is the bimetallic temperature-

Clothing Iron Troubleshooting Chart

Trouble	Probable Cause	Remedy
No heat.	Line cord is defective.	Check line cord for continuity. If a replacement is necessary, be sure to use heavy-gauge asbestos insulated cord, often called heater cord.
	Thermostat contacts are stuck open.	Check thermostat. Free contacts, if possible, or replace thermostat.
	Heating element is defective.	Check heating element for continuity. Replace element or soleplate as needed.
Excessive heat or erratic heat control.	Thermostat contacts are short-circuited.	Check thermostat for short circuit.
	Thermostat needs to be recalibrated.	Use iron tester—if available—to recalibrate thermostat.
	Short-circuited heating element.	Check heating element for short circuits.
Iron will not shut off.	Thermostat is defective.	Check thermostat; replace if defective.
Steam feature or spray feature does not operate.	Defective or dirty parts in steam or spray feature.	Inspect steam chamber, valves, tubes, and other parts. Clean out clogged ports, openings, tubes, and valves. Replace parts as needed. When replacing parts, install new gaskets.
Steam chamber or water tank leaks into iron.	Corrosion or expansion and contraction has caused leak.	Replace affected parts and install new gaskets.
Iron spits.	Thermostat set too low.	Set thermostat to higher temperature.
	Corroded parts.	Clean iron.
	Water tank too full.	Partially empty water tank.
Iron stains clothing.	Starch on soleplate.	Clean soleplate.
	Minerals or sediment in tank.	Clean tank with solution of ½ vinegar and ½ water; use distilled water when ironing.
Steam ports or spray nozzle openings are clogged.	Mineral deposits or sediment is clogging ports, openings, tubes, and valves.	From the outside of the iron, clean the ports and openings with thin wire or a stiff brush bristle.
Soleplate scratched or stained.	Improper use; excessive use of starch.	Polish on buffing wheel or some equivalent, using emery compound for heavy-duty work and jeweler's rouge for a high-gloss finish.
Iron blows fuse or trips circuit breaker.	Short circuit.	Check line cord; then make visual inspection of heating element, thermostat, and wiring. Next, check for continuity until short circuit is found. Replace defective parts.
Iron shocks user.	Short circuit to body of iron.	First inspect visually. Then make continuity tests until short circuit is found; replace defective parts.

sensitive element. It is usually made of brass and iron and has a porcelain insulator on it that presses against either the upper arm or the lower arm, depending upon the manufacturer's design. When the temperature lever is set on the iron, it sets the spacing between the upper arm electrical switch contact and the lower arm electrical switch contact. As the iron heats and cools, the bimetallic temperature-sensitive element bends forward and backward, opening and closing the electrical switch contacts as required to maintain the temperature selected by the temperature lever. The thermostat is set against the soleplate and the silver or silver-plated brass contacts open and close with about a 5 degree F change in temperature.

Several kinds of problems can occur in the thermostats in irons. The bimetallic blade may break, or the contacts may become permanently opened or closed. The contacts may also become pitted, worn, or stuck together. Pitted contacts cause additional heating at the contacts, which results in less heat at the element, thereby causing temperature-calibration problems. Leaks from the water tank of a steam iron can cause corrosion and subsequent breakage. No matter what the problem is with the thermostat, it is recommended that the thermostat be replaced with an identical part. Generally, individual parts for thermostats are not available.

When a steam iron user desires steam from the iron, he presses the jet stream control, which pushes against a plunger and opens a valve, letting water into a steam chamber, where it turns into steam. The spray control pushes against a mechanism that causes water from the water tank to be sprayed through the spray nozzle onto the fabric being ironed.

Take note in regard to line cords that if an iron electrical cord needs replacement, it must be replaced with the same type of special cord having the same or larger wire gauge. Iron cords are asbestos-insulated to protect the cord against the high temperatures of the iron. One end of the cord contains a regular electrical plug; the other end has special solderless eyelets or lugs that mate with the electrical terminals inside the iron. A cord sleeve made of molded rubber or neoprene is located at the point where the cord wires enter the iron to help in preventing breaks in the cord because of the constant back and forth motion of the iron. Despite this feature, faulty cords are the most prevalent malfunction in an iron.

Some of the major causes of trouble in a steam iron are caused by incorrect use of the iron. Tap water containing minerals and impurities is often used instead of distilled or demineralized water. Also, the user sometimes fails to empty the tank before putting the iron away. By using regular tap water, the impurities and minerals from the water cause the valves and vent holes to clog. The tank may get sediment in the bottom from the impurities; the tank may also leak. Other problems that can occur in the steam part of steam irons include leaking gaskets and bent control rods. Leaky gaskets should always be replaced with new ones. Control rods can sometimes be straightened out; but if they cannot be, they must be replaced.

Other common malfunctions in automatic dry irons and automatic steam irons are electrical malfunctions that cause either no heat, excessive heat, or erratic heat for the setting of the temperature lever. Common mechanical malfunctions usually involve some failure of the steam or spray features or a water leak inside the iron. Of course, these are in addition to the already mentioned prevalence of line cord problems encountered with irons.

Servicing techniques for irons are all generally the same. However, for specific instructions, such as disassembly and reassembly of complex irons and for the calibration of thermostats, it is suggested that you refer to the manufacturer's instructions whenever possible. When making electrical continuity tests, be sure that the clothing iron plug is removed from the convenience outlet. Set the temperature lever to linen to ensure that the thermostat contacts are closed. Connect the VOM probes between the line-cord plug prongs. The VOM should indicate between about 10 and 25 ohms; less than 10 ohms indicates a short circuit within the iron. An infinite resistance indication tells you there is an open circuit in either the thermostat or the heating element. Make further continuity tests as necessary to isolate the trouble to either the thermostat or the heating element. Also check that there is not a grounded condition between the electrical circuit and the iron case.

It is necessary for the iron to be at the proper temperature for the fabric being ironed; it must also be at the correct temperature to produce the proper steam pattern. Whenever it is suspected that the thermostat is not controlling the iron temperature within the proper range or when a thermostat is replaced in an iron, thermostat tests using an iron tester should be made to check and/or set the calibration of the thermostat. In checking the range or proper operation of the thermostat, check it on both ends of the temperature range, such as rayon on the low end and linen on the high end. To use a tester properly, place the iron soleplate on the tester and plug the iron cord plug into the tester appliance power receptacle. Set the iron temperature lever to the specified range using the manufacturer's suggested procedure. Observe the cycling indicator light on the tester. Take a temperature reading after three ON-OFF cycles; then after several more cycles, take two or three additional readings, which should be nearly consistent. If the readings are erratic, the thermostat should be replaced. If you are calibrating the thermostat, you can select a midrange such as that for wool. Allow the iron to heat up and cycle on and off at least three times. If necessary, adjust the thermostat setscrew either clockwise or counterclockwise to bring the iron within the proper temperature range.

Thermostat Temperature Ranges for Irons

Fabric	Temperature (°F)
Wash and wear	200 ± 20
Rayon	245 ± 50
Silk	310 ± 50
Steam position	350 ± 40
Wool	395 ± 50
Cotton	460 ± 50
Linen	510 ± 50

If it is necessary to disassemble an iron for repair, follow the manufacturer's instructions, if available. Steam irons are generally rather complex, and you may need special tools to hold certain parts during disassembly and reassembly. Remember, do not disassemble more parts than are necessary. Any connections within an iron that require soldering must be soldered using silver solder, the correct flux, and a propane torch. This silver soldering or brazing, as it is called, must be able to withstand temperatures of 1,000 degrees F when completed.

Steam iron ports, nozzle orifices, and tubes often become clogged with dirt, sediment, or mineral impurities. From the outside of the iron, this debris can be cleaned out as much as possible with a paper clip or some other thin wire, a stiff bristle from a brush, a needle, or a small drill bit. Valves may be cleaned in the same manner. Excessive mineral deposits within a steam iron are cleaned out by making a solution of ½ cup of white vinegar and ½ cup of water. Pour the solution into the steam iron tank and let all of the solution steam through the iron. The vinegar dissolves the minerals and the residue is steamed out of the iron.

Soleplates can be polished with an emery compound on a buffing wheel to remove heavy scratches and stains. A high gloss finish can be produced on the buffing wheel, using jeweler's rouge. Dirt on the rest of the iron can be removed by using a household detergent.

Electric Blankets and Heating Pads

An electric blanket is used to keep you warm in bed. What the blanket actually does is to maintain your body temperature at a normal 98.6 degrees F. On the other hand, a heating pad raises the temperature of a part of your body above normal. Therefore, an electric blanket operates according to the room temperature.

Electric Blankets A very common complaint is that an electric blanket does not get as warm as it should. Often there is no blanket malfunction; the problem is a failure to fully understand the function of an electric blanket. As already mentioned, the blanket is designed to keep the body at normal temperature. If the room temperature is 75 degrees F, the electric blanket may not come on at all except at perhaps the very highest control setting. The control box should sit on the floor or on a table near the bed. It must not be placed in a position near a radiator, an air duct, a window, or an open door. It should also not be located where there is a draft or where it is not in the typical room environment. If it is located incorrectly, you will get the false impression that the blanket is operating incorrectly.

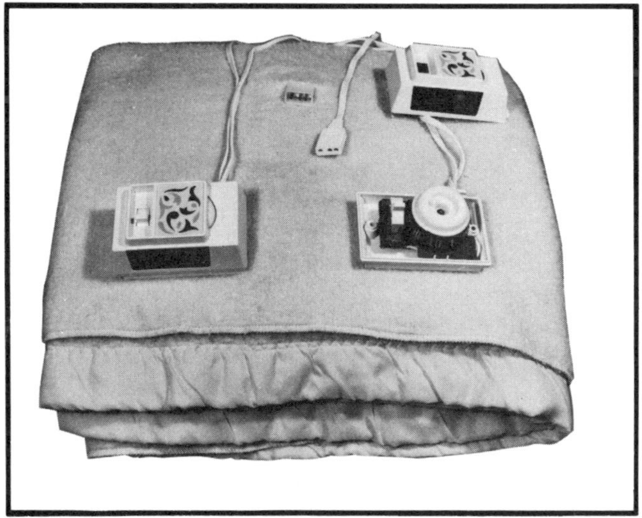

Electric blankets and heating pads are similar in design and electrical configuration; however, an electric blanket is only supposed to maintain the body's temperature while a heating pad is designed to raise it in limited areas.

Electric blanket circuits are relatively easy to understand. The electric blanket consists of two flexible heating elements—sometimes there may be only one—a thermostat control, safety thermostats, and a two-wire cord from the plug to the control box with a three-wire cord from the box to the blanket. The flexible heating elements are thin-stranded wire covered with a fireproof covering. Generally,

Electric Blanket and Heating Pad Troubleshooting Chart

Trouble	Probable Cause	Remedy
Complete blanket or section of blanket does not heat.		Isolate problem to either blanket or control box by making continuity tests on blanket.
	Line cord is defective.	Check for continuity. Replace if necessary.
	ON-OFF switch is defective.	Check for continuity. Replace switch if necessary.
	Thermostat is defective or dirty.	Clean thermostat contacts; then check thermostat with VOM. Replace if necessary.
	Heating element or safety thermostat in blanket is defective.	Check for continuity with VOM; if test indicates infinity in one or both sections, replace blanket.
Blanket does not cycle on and off.	Thermostat is defective.	Rotate thermostat through range. With control box located in area where temperature is below 68 degrees F, the blanket should come on. Check thermostat with VOM.
Blanket gives insufficient warmth.	Temperature dial set too low.	Increase warmth by changing setting on dial.
	Thermostat is defective.	Rotate thermostat through range. With control box located in area where temperature is below 68 degrees F, the blanket should come on. Check thermostat with VOM.
Blanket gets too hot.	Thermostat is not cycling.	Check thermostat. Replace if necessary.
Indicator light does not light, but blanket works.	Light is defective.	Check light circuit for continuity. Replace as needed.
Person receives slight shock from blanket.	Contacting a sensitive portion of the body to a grounded object—the blanket—while insulated by the blanket.	Plug into different convenience outlets; if two plugs, reverse their positions.
Appliance blows fuse or trips circuit breaker.	Short circuit.	Test for short circuits within control box or boxes and in electric blanket. If short circuit is in blanket, discard it.

one element is located on each side of the blanket, where each is sewn between two pieces of material. A three-prong connector mating with the heating elements attaches at the center bottom of the blanket. All the other components are contained within the control box or boxes. A blanket that fits a twin-sized bed has one control; a blanket that fits a double-sized, queen-sized, or king-sized bed may have only one control, but in most models there are two controls. There is a separate thermostat control in each control box so that the two sides of the bed are individually controlled.

Several safety thermostats are located in series with and along the length of the heating elements within the blanket. If the blanket is folded over or if a heavy object is placed on top of it, causing the temperature within a small area to rise to a level that could eventually become unsafe, the normally closed contacts of one of the safety thermostats open, causing the current flow in the series circuit to stop flowing. There may be up to 10 safety thermostats within a blanket.

Usually it is impractical to repair a malfunction within the blanket itself. It is very difficult to locate an open circuit in an element in a blanket; and to repair it would mean opening the blanket material, locating the trouble, repairing it, and resewing the blanket. However, you can test the elements in the blanket very easily by making a continuity check with the control circuit disconnected from the blanket. A typical example of the value of each element for a double-sized bed is approximately 150 ohms. Since one side of each element is connected together within the blanket, it is possible to measure the resistance of the blanket from end to end. The element would then be in series for a continuity test. The total resistance would be approximately 300 ohms. Always be sure that the electric blanket control box power plug is disconnected from the convenience outlet before making a continuity test. Troubleshooting is performed to isolate any trouble to the control box or to the blanket. Symptoms of a malfunctioning blanket are failure to heat, insufficient heat, or too much heat. The continuity check, as previously mentioned, can help you to isolate the malfunction to the control box; then you can make continuity tests within the control box to isolate the malfunctioning component.

Heating Pads A heating pad is similar to an electric blanket; however, a heating pad is used to raise the temperature in a certain part of the body to a temperature above the normal body temperature. The heating pad is not designed just to keep you warm. Like the electric blanket, there are heating elements sewn into the heating pad; usually there are two. The pad is rubberized to make it more flexible for applying the heat to that part of the body, such as the back muscles, where the heat is needed. The control unit is contained in a small control box that is separate from the pad. There are two wires in the input cord to the control box that carry the 120 volt AC power. Three wires lead out of the control box to the heating pad. One of the three wires is common to both heating elements and the other two wires are for the dual heating element arrangement, one element with about 20 watts, the other about 40 watts of heat.

There are two types of heating pad configurations; one is not thermostatically controlled and the other one is. On a heating pad that does not have thermostatic control, the control box has a latching or rotary switch control circuit with Off-Low-Medium-High switch positions. The heating pad itself has two heaters with safety thermostats connected in series with each of the heating elements. In the event that the temperature of the heating pad rises above the temperature that is considered safe, the safety thermostat contacts open, breaking the path of current flow and turning off the heating element.

The second type of heating pad is one that is controlled by a thermostat. The control box in this circuit contains a control thermostat, a neon light, switching system, and two bias thermostat heaters. The bias thermostat heaters operate in conjunction with the control thermostat to regulate the heat in the heating pad. The heating pad itself contains one heating element and an overheat safety thermostat connected in series with the heating element. If the pad becomes too hot, the contacts of the overheat safety thermostat open, breaking the path of current.

Heating pads operate at about 50 to 60 watts of power. The power can be monitored on a wattmeter, an ammeter, or with some VOM's and should be within ± 10 percent of the reading on the nameplate. A thermocouple—a thermoelectric sensor used to measure temperature differences—test can be made to check the temperatures of the heating pad. Place the thermocouple under the pad. At the low setting, the temperature should be between 100 and 130 degrees F; at medium, between 120 and 150 degrees F; and at high, between 140 and 180 degrees F.

Large Appliances

Large appliances have been put into a separate category, not only because they are larger and generally more complex than smaller ones, but also because they are harder to replace if they do break down. If your blender malfunctions, you can always whip cream by hand or do any of the other jobs done by a blender without wasting much time. However, replacing a defective oven or refrigerator is well nigh impossible, short of buying a new one, and having to wash clothing by hand will take a great deal of time and work. Homeowners are usually much more dependent on their larger appliances; therefore, when they break down, the cost to the homeowner in terms of time and energy is much greater.

Nevertheless, you cannot let yourself be intimidated by the thought of repairing a large appliance. True, they are more complex, they have a greater number of parts, more complicated electrical circuits, and so on. But, except for the introduction of a few parts not used in small appliances—special timers, refrigeration systems, and humidistats, for example—large appliances are not really that different. The same basic electrical rules, circuitry, and mechanical rules apply to all appliances. The larger appliances will often require more voltage, but as long as you have developed safe work habits, this is not a matter of concern.

Refrigerators

The household refrigerator can be defined as an automatically controlled, insulated cabinet, shaped and arranged for the storage of perishable foods. Heat is removed from the interior of a refrigerator by mechanical means to produce a storage temperature most satisfactory for food preservation and to provide a supply of ice cubes. Most home refrigeration units manufactured today are combination units; they contain two separate compartments, the refrigerator and the freezer sections. Each of these compartments is heavily insulated from the other, and each operates at a different temperature. The provision or refrigerator section, for instance, is usually maintained at a user-selected temperature which may range from 34 to 42 degrees F, while the temperature in the freezer section is normally kept at approximately 0 degrees F. The freezer compartment may be either above or below the refrigerator or provision section; some larger home refrigeration units have a side-by-side arrangement.

Home refrigerator units have four basic areas with which you must be familiar. You must also be aware of the different problems which are common to each of these areas. They are the cabinet, the air-handling system, the electric and control system, and the refrigeration system.

Cabinets The cabinet is the housing for the space to be refrigerated. It also houses the self-contained refrigerating system. That part of the cabinet surrounding the food compartment is heavily insulated to retard the flow of heat into the compartment. The door to the food compartment is well insulated and must be made to close tightly. Cabinets of refrigerators consist primarily of the outer case, inner liner, and insulation. The outer case is steel finished with baked enamel. Outer cases as a rule are not available for replacement.

A typical refrigerator-freezer combination with the freezer unit on top.

The liner on most models is steel with welded or crimped seams and is finished with porcelain or baked enamel. Some models, however, have molded plastic liners. Many combination refrigerators have a separate liner for each compartment, but some later combination models use a single liner with a dividing partition between the freezer and fresh food compartments. Many models have fiberglass and foam slab insulation, and the liner is fastened to the outer case with screws through nylon spacers and plastic or ceramic supports at the corners. Some models use foam insulation poured between the outer case and liner that bonds the two together. On these models, the liner is usually not available for replacement.

Complete disassembly of the cabinet is required if the liner or insulation is replaced. Remove doors, drawers, baskets, shelves, bins, and all other lift-out parts. When replacing the liner, transfer any brackets, braces, nut strips, or supports not provided with the replacement part. When servicing the outer case of any refrigerator, be sure in reassembly to use permagum or a similar material as a sealer around any screws or fasteners which pass through holes in the outer case. This will ensure minimum possible leakage of air into the insulation.

Refrigerator doors consist primarily of an outer door panel, an inner door panel, insulation, and a gasket. The inner door panel is usually held to the outer door panel with screws located under the gasket flange. Retainer strips, when used, hold the door gasket and add stiffness to the edges of the inner door panel. While fiberglass is still the most popular door insulation, some models use a combination of fiberglass and foam slabs. Many larger models now use foam insulation poured into the outer door panel. The door swings on heavy hinges and is held closed by a latch mechanism. A door gasket, fitted around the periphery of the door's inner panel, seals against the cabinet when the door is closed and prevents the infiltration of air into the food compartment. Actually, maintenance of the proper storage temperature is dependent upon the efficiency of this gasket seal. Air leakage will also cause a rapid buildup of frost, condensation, and longer operating cycles.

Air-Handling Systems Air circulation is important both inside and outside the refrigerator. Outside, the heat produced by the compressor operation sets up a hot zone around the lower part of the refrigerator cabinet; therefore, air circulation is created in the form of an escape current rising up the rear part of the refrigerator and an intake draft passing from the lower front part. It is essential that this draft not be interfered with. For this reason it is necessary for the refrigerator to be spaced away from the wall. Spacers are always attached to the condenser to ensure that the necessary amount of room is provided.

Inside the refrigerator, natural airflow in a non-frost-free or a non-fan-cooled conventional system is by convection. There are two zones within the refrigerator: the evaporator—the freezer area—and the provision or general storage compartment, each having distinctly different temperatures. There are, in fact, air currents passing between the top and the bottom of the storage compartment because of this temperature difference, and these must be taken into consideration to get the best distribution of the cold in the storage compartment. The relatively warm air of the lower part of the liner rises to the upper zone via the front space, between the door compartments and the shelves. In return, a current of cold air descends from the evaporator to the lower zone of the liner between the shelves and the liner.

A refrigerator using gravity circulation has its evaporator or cooling coil located near the top of the compartment. Warm air rises and cool air drops. As the air is cooled by the evaporator, it drops to the lower section of the refrigerator. The warmer air rises up to the evaporator where it is cooled and again drops. This sets up a circulation of air in the refrigerator. As it circulates, it picks up heat from the food and deposits it on the evaporator. If anything blocks the natural airflow, the food-cooling action of the refrigerator is reduced. This is the reason open wire shelves are used instead of solid shelves. An overly fussy homeowner who places paper on the shelves will soon find the food in the lower area of the refrigerator spoiling, while food in the upper area near the evaporator will be too cold or even freezing. Packing the food too closely together so that air cannot circulate will also cause the same problem.

In the typical self-defrosting, forced-draft top-mounted refrigerator-freezer, the air in the freezer section is pulled into the front grille and over the entire width of the cooling coil. Heat and moisture are given up to the coil with the result that cold, dry air is discharged out of the air duct into the freezer compartment. An air deflector is used to divert air directly over the ice service area. The fan or blower housing is designed so that some of the air is diverted to the food compartment. The position of the modulating air regulator or manual damper will determine the amount of air which flows into the food compartment. The air from the food compartment returns to the freezer section through the slots in the divider located on each side of the control housing. In the freezer it mixes with the freezer return air and passes through the evaporator and back to the fan or blower motor

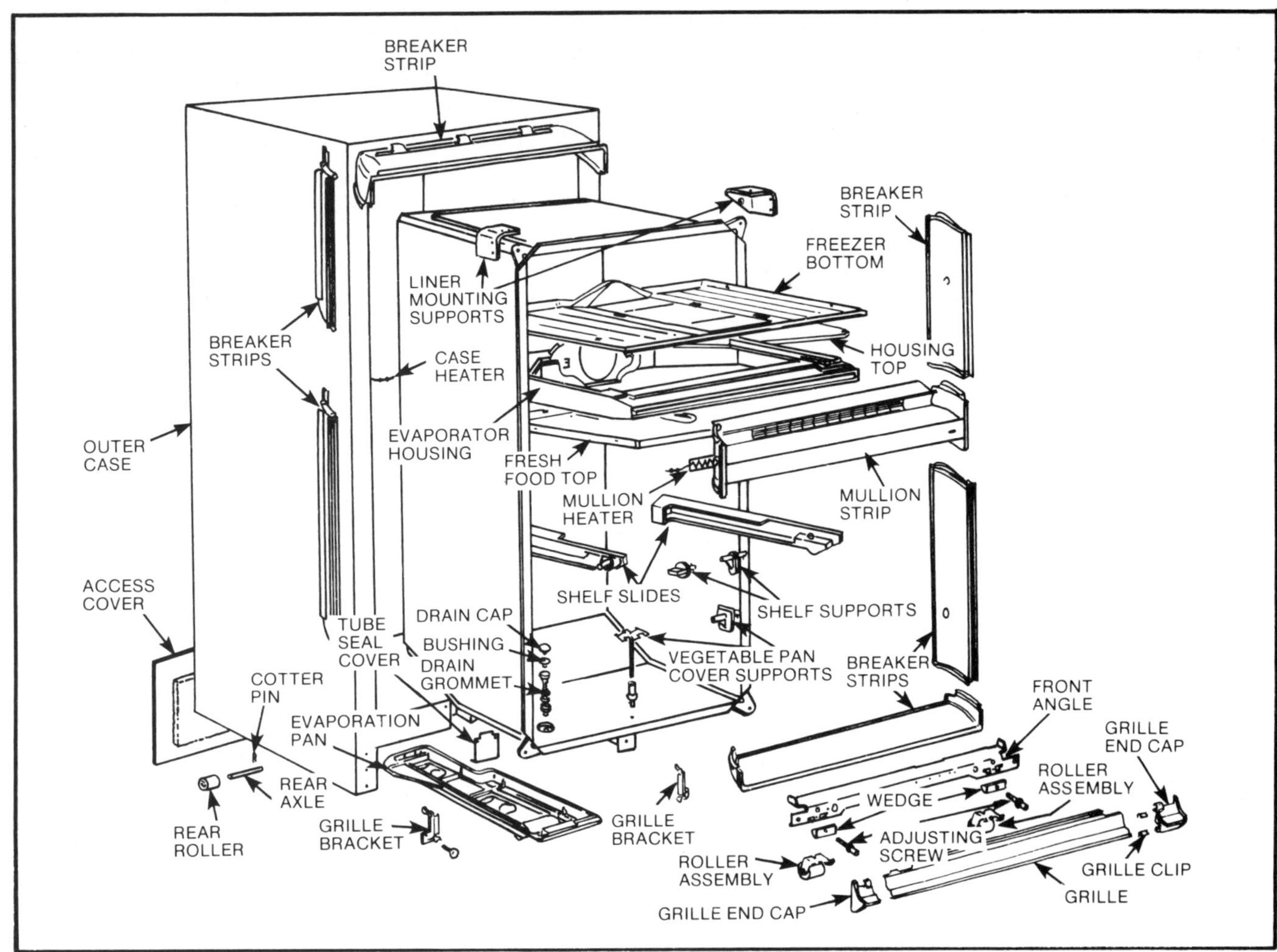

Exploded view of a typical refrigerator cabinet assembly.

housing. The defrost water is removed from the freezer compartment through the drain tube, then down the rear wall of the food compartment to the drain opening in the bottom of the food compartment liner. Defrost water then goes through a drain trap into a disposal pan in the machine compartment, where it evaporates. The drain trap should be cleaned periodically. The forced-air system in a bottom-mount refrigerator-freezer is usually the same as in the top-mount type.

Electrical and Control Systems There are two basic electric circuits in a refrigerator: the circuit that operates and controls the cooling cycle and the one that does the same for the defrost cycle. The cooling circuit includes the cold control which is the thermostat, current relay, overload protector, and motor-compressor. The function of the cold control is to sense and control temperatures by starting and stopping the motor-compressor. The current relay energizes and de-energizes the compressor start winding. The overload protects the compressor against high amperage and high temperatures, and the motor-compressor, of course, compresses and circulates the refrigerant.

A refrigerator has a defrost system because the evaporator collects moisture in the form of frost. This accumulation of moisture must be periodically removed to maintain a proper cooling temperature. If frost is allowed to build up, the air moving over the fins of the evaporator will be blocked, no heat transfer will take place, and the refrigerator and freezer temperatures will rise. To remove this frost, the refrigerator must be defrosted. This procedure can be accomplished automatically or manually, depending on the model refrigerator. Of course, if a refrigerator requires manual defrosting, it will not have a separate defrost circuit.

The most popular method of automatically defrosting a refrigerator is the timed-defrost method employing heating coils. The four basic components in this type of defrost circuit are the defrost timer, the sheath or defrost heater, the defrost thermostat or limiter, and the drip-trough heater. The automatic defrost timer determines or initiates when the refrigerator goes into its defrost cycle. The sheath or defrost heater is the electric heating element which defrosts the evaporator. The defrost thermostat or limiter controls the length of time heat is applied to the evaporator. It is mounted in a location that allows it to sense the evaporator temperature, thus indicating when all the frost is removed. Most defrost thermostats turn off when the evaporator temperature reaches approximately 55 to 70 degrees F. The defrost heater cannot be energized again until the defrost thermostat bimetal limit switch has been cooled down to its resetting temperature at about 30 degrees F. The drip-trough heater is generally the blanket type sealed to aluminum foil and wrapped around the underside of the drip trough. The trough heater prevents defrost water from freezing in the drip trough and possibly blocking the drain and airflow to the evaporator. Some models employ a radiant defrost heating element, in which case a drip-trough heater is not necessary. The radiant heating element is below the evaporator and serves as both the evaporator and drip-trough heater.

Refrigeration System The refrigerant cycle consists of heat-laden refrigerant being pumped from the compressor to the condenser on the cabinet back. In the condenser, the heat is passed off into the surrounding room and the refrigerant is condensed into a liquid. The liquid refrigerant then passes through the capillary tube to the evaporator. The refrigerant enters the larger evaporator tubing, which releases the pressure on it. This allows the refrigerant to boil and absorb the heat from the cabinet. The heat-laden vapor returns from the evaporator via the suction tube.

To be able to successfully diagnose a problem in a defective refrigeration system, it is important to understand the characteristics of a normally operating system. A correct charge of refrigerant is indicated by a normal wattage draw and temperature differential as specified in the performance chart for the particular model. When the refrigerant charge is normal, most of the refrigerant will be in the evaporator during the cooling cycle, and only the final few passes of the condenser will contain liquid refrigerant. Approximately 90 percent of the condenser will contain high-pressure gas in the process of heat dissipation and condensation.

About 3 minutes on the OFF cycle is required for the high- and low-side pressure to equalize and thus permit the compressor to start at the beginning of the next run cycle. The temperature of the discharge line falls during the OFF cycle. At the beginning of the run cycle, there is a noticeable decrease of the suction line temperature and an increase of the discharge line temperature.

Repairs and Servicing Most refrigerator problems are electrical or related to the air circulation system. Refrigeration system failure is rare. Line cord failure is also rare because the cord suffers little or no abuse.

An open thermostat will keep your refrigerator from working. To check it, disconnect at least one lead and test across the terminals for continuity. To be sure the sensing bulb is warm enough to close the contacts, hold it in your hand. Also, if the thermostat sensing bulb has come loose from its mount for any reason, the compressor will run too long and the temperature inside the refrigerator will be too low. A visual inspection will reveal a loose sensing bulb, which should be easy to remount.

An open compressor overload protector, open relay coil, or open motor windings will, of course, also shut down the system. Generally, you will find the overload protector and relay mounted on top of the compressor housing. The protector is the smaller of the two and is usually cylindrical in shape. Both the overload protector and the relay can be easily checked for continuity with the VOM. Test for open motor run windings across terminals usually marked C for common and R for run. The start windings are checked for continuity across the terminals usually marked C and S. If you find open motor windings you will have to replace the compressor unit.

One of the most common of all refrigerator malfunctions is a leak in the door seal. A refrigerator door receives a lot of abuse; therefore, the gaskets are frequently worn or damaged. A leak in the door seal will cause the compressor to work constantly; however, you can buy replacement gaskets and they are easy to install. It is best to start remounting at one point; then move along the gasket, tucking and tightening until you have completed the job.

A warm refrigerator compartment accompanied by what appears to be a normally operating refrigerator is often the result of evaporator icing. If you have a cycle defrost model, remove the evaporator cover panel and look at the frost pattern. An even pattern indicates that the refrigeration system is working properly. A defective thermostat might be cycling the compressor on before the defrost cycle has had a chance to completely melt the frost away. A defective thermostat might also be keeping the compressor running longer than necessary for the setting. A partially frosted evaporator accompanied by a constantly running compressor indicates a leak in the refrigeration system. An inadequate refrigerant supply will make the thermostat act as if the refrigerator is warm, again causing the compressor to overwork.

In the defrost system of a self-defrosting refrigerator, a defective defrost timer will cause evaporator icing. The accumulated ice will be white and snowy looking, but hard as a rock. If the drip-trough heater becomes defective, water draining from the evaporator will freeze and eventually back up to the coils. Water on the refrigerator floor indicates a clogged drain.

Refrigerator Troubleshooting Chart

Trouble	Probable Cause	Remedy
Refrigerator does not work at all.	Line cord is defective.	Check voltage at relay. If there is none but voltage is indicated at outlet, replace line cord or plug on cord.
	Loose electric connection or broken lead.	Check components for continuity with VOM. Repair defective part.
	Air circulation is unsatisfactory or compressor is overheated.	Clean condenser. Check condenser fan; replace if defective.
	Cold control or thermostat is defective.	Check for continuity; replace if defective.
	Relay and overload not working.	Use starting cord to check compressor directly. If compressor starts, use VOM to check relay and overload separately. If compressor does not start, replace compressor.
	Start capacitor is defective.	Check capacitor for continuity; replace if defective.
	Compressor motor is defective.	Check motor for short or continuity; if motor is defective, replace compressor.
	Defrost timer is defective.	Check; replace timer if necessary.
Compressor motor runs but does not refrigerate or refrigerates poorly.	Moisture restriction; heavy frost around evaporator inlet.	Heat frosted area. If frost line moves farther along coil after heating, restriction was probably caused by a moisture free-up. Discharge the unit; and recharge.
	Permanent restriction; pinched tubing.	First check for moisture restriction; then check for crimped or damaged tubing. Repair or replace restricted part.
	Low charge or no charge of refrigerant.	Check for leak; repair leak. Then purge and recharge unit.
	Air circulation is not flowing properly.	Do not cram too much food into refrigerator, blocking air currents.
	Condenser or grille restricted by lint.	Clean condenser and air passage with a vacuum cleaner.
	Condenser fan not working or running very slowly.	Check fan motor; replace if defective.
	Lower air baffle is missing.	Replace air baffle.
Refrigerator runs excessively.	Normal summertime operation.	—
	Placing full load in refrigerator.	—
	Making heavy use of the refrigerator; keeping it filled; opening the door frequently.	Open the door as little as possible.
	Gasket seal is poor.	Adjust door hinges and check to see if seal is tight; if worn, replace gasket.

Refrigerator Troubleshooting Chart (Continued)

Trouble	Probable Cause	Remedy
Refrigerator runs excessively.	Control is at too cold a setting.	Lower temperature setting.
	Airflow around condenser restricted or bypassing condenser.	Adequate airflow over condenser is necessary.
	Restriction or moisture.	Replace component where restriction is located. If moisture is suspected, replace dryer-filter.
	Frost coil on automatic defrost models is defective.	If air cannot get through to switch bulb, unit will not cycle.
	Obstruction in air duct.	If inlet or outlet duct has an obstruction or frost accumulations, air cannot actuate switch bulb. Remove obstruction.
	Condenser fan is malfunctioning.	Check for fan motor operation and obstruction of fan blade.
	Interior light burns constantly, raising temperature.	Check light switch; replace if defective.
Refrigerator and/or freezer are too cold.	Control knob set too cold.	Reset cold control/thermostat to a warmer setting.
	Thermostat contacts stuck closed or thermostat is otherwise malfunctioning.	Check thermostat; replace if defective.
	Thermostat bulb connection is loose.	Check connection and make necessary adjustments.
	Control capillary is loose.	Retighten clamp.
Refrigerator and freezer are too warm.	Thermostat is defective.	Replace thermostat.
	Excessive frost on evaporator.	Check defrost timer, defrost thermostat, and defrost heaters; replace defective parts.
	Excessive door openings and food loads.	Open the doors as little as possible, and do not cram the compartments completely full.
	Door seal is not tight.	Check door gasket; replace if defective.
	Evaporator fan is defective.	Check fan motor and fan switch; replace defective parts.
Refrigerator is too warm, but freezer temperature is normal.	Freezer cold control improperly set.	Set freezer control to a lower number to permit more air to be transferred to the food section.
	Loose air seal gaskets.	Check that all gaskets in the air transfer from the freezer to the fresh-food section are firmly in position to prevent air leakage.
	Fresh-food air return path is blocked.	Make sure that the air return passage is not blocked by an ice buildup. Check all heaters used to prevent ice buildup in the air duct systems.
	Freezer fan motor or fan switch is defective.	Check fan motor and fan switch; replace if necessary.

Refrigerator Troubleshooting Chart (Continued)

Trouble	Probable Cause	Remedy
Refrigerator is too warm, but freezer temperature is normal.	Stalled fan.	Make sure fan is not blocked by ice; check heaters.
	Malfunctioning defrost system.	Check defrost system; make sure freezer coil is defrosting.
	Evaporator is heavily frosted.	Needs defrosting more often; also, check door gaskets for air leak.
	Food compartment control thermostat is defective.	Make sure thermostat opens food compartment duct.
Inside of refrigerator cabinet sweats heavily.	Storage habits not helpful to a proper refrigeration operation.	Cover all foods and liquids. On manual defrost models, regular defrosting is necessary.
	Door seal is not tight.	Check door seal; make necessary adjustments or replace gaskets.
	Cabinet seal is bad.	Wet insulation is usually a sign of a bad cabinet seal. If a bad cabinet seal is suspected, remove liner, unit, and insulation, and reseal cabinet seams.
	Line bottom heater is defective.	Condensation on bottom of refrigerator compartment could be result of inoperative line bottom heater. Check heater with VOM.
Incomplete defrosting.	Limit switch is defective.	Check bimetal limit switch usually located near left side of evaporator coil; replace if defective.
	Defrost timer is defective.	Check defrost timer; replace if defective.
	Heaters are defective.	Check drain heater, shroud heater, and defrost heater; replace malfunctioning heaters.
Odor in refrigerator.	Odorous food.	Cover odorous foods when they are stored in the refrigerator. If the odor lingers after foods have been removed, clean the refrigerator and rinse with a solution of baking soda and water. You can also leave an opened box of baking soda in the refrigerator.
	Drain system is dirty.	Clean drain system and flush with a solution of baking soda and water.
	Filter, if used, is too old or dirty.	The effective life of a filter is about one year. If filter has been in use for more than one year, replace it.
Refrigerator blows fuse or trips circuit breaker.	Short circuit.	Check for short circuit; replace shorted parts.

Finally, as with most appliances, especially the large ones, a little bit of preventive maintenance and some forethought about the refrigerator's operation can save you a lot of grief. Do not overload a refrigerator with warm food, make a point of keeping the condenser coils clean, and do not leave the refrigerator door open any longer than is necessary. Obviously, your refrigerator will last a lot longer if it is not forced to do a lot of unnecessary work. Also, make sure the door seal is tight without the possibility of air leakage.

Room Air Conditioners

A properly designed room air conditioner should perform five separate tasks. First and most obvious, the room air conditioner should regulate the temperature in a room. Second, it should regulate and control humidity. It should also furnish proper ventilation, recirculate the contained air within the conditioned space, and it should filter and clean the air. In reality, there are only two basic systems in a room air conditioner: the air circulation and the refrigeration systems.

There are two kinds of air circulation systems in a room air conditioner. The air system on the outside includes the condenser fan and dissipates the heat taken from the cooled area to the outside atmosphere. The condenser fan also picks up and dissipates into the outside atmosphere the water which has been removed from the air in the cooled area. The slinger ring on the outer periphery of the fan performs this function.

In the second type of air circulation system which operates in the cooled area, room air is circulated over the cooling unit. In so doing, the air loses heat to the cooling unit. The cooling unit also condenses water vapor from the room air. A solid insulated partition or bulkhead inside the air conditioner separates the outside and inside air systems. Therefore, these systems are so arranged that the cooling coils which absorb heat are located on the room side of the window and the condensing unit which rejects heat is on the outside of the window. It is important that no intermingling of the hot condenser air and cooled room air takes place. This is the primary purpose of the partition or bulkhead; it separates the room air from the condenser air. The rubber seals and panels used at the window prevent the exchange of inside and outside air. Needless to say, tight installation at the window is of utmost importance.

Fan System The fan system is not extremely complex. The evaporator fan and condenser fan blades attach to each end of the fan motor. As a rule, a squirrel-cage type evaporator fan is employed because it has a low sound level and a high air-pulling capacity. It must pull the room air through the evaporator coil and expel the cooled dry air back into the room. A blade-type fan may be found on some small compact conditioners where blower space is limited. On the outside end of the motor shaft is the condenser fan. This will almost always be a blade-type fan which blows the heated air outside.

If you suspect trouble with the blades or fan motor, be sure to set the switch in the OFF position before examining either the evaporator fan or condenser fan blades. Rotate the blades slowly by hand and listen for a scraping or grating noise. If blades are hard to turn, or if you feel or hear any scraping or grating noise, the air conditioner chassis must be removed from the cabinet to permit a complete examination to determine the cause of the trouble and its proper correction. Occasionally a fan blade setscrew becomes loosened and emits a clicking sound between the fan blade hub and the motor shaft. Loosened setscrews can be detected by rocking the blades of the evaporator fan. As the fan blades and motor shaft assembly change direction of rotation, the loosened setscrew will bump on the shaft. Before a loosened setscrew can be tightened, the air conditioner chassis must be removed from the cabinet.

The fan motor can be either a single-speed or multispeed type, depending on the model. It can also be either a shaded-pole or a permanent split capacitor type. All quality room air conditioners are equipped with built-in overload switches. These switches may not be shown on wiring diagrams because of space limitation, but have always been in use.

A typical room air conditioner unit.

If you have a problem with a fan motor, make sure that all electric connections between the power source and the fan motor are good. Most fan motors are also equipped with a capacitor which should be tested. To test the fan motor windings, remove the motor from the terminal connections and check the continuity of the windings with a test light or the VOM. An open circuit indicates a defective winding, a broken connection, or a bad motor protector inside the motor. Continuity between the windings and the frame indicates an internal short or ground, and the motor must be repaired or replaced.

Many fan motors are equipped with oiler tubes. An SAE-20 weight nondetergent type of oil is recommended. A detergent oil can ruin the oil-absorbing quality of the bearings used in these motors. If the motor does not have oiler tubes, it has a lasting lubricant and therefore requires little attention. Do not over-oil as the reservoir may overflow and damage the motor. Be sure to replace the oiler plugs after oiling to keep water and dirt from getting into the fan motor bearings.

Refrigeration System In this process, the refrigerant used in the air conditioner is changed from a liquid to a vapor and then restored to a liquid. The heat energy changing the liquid refrigerant to a vapor is the heat extracted from the air in the conditioned space. The complete cycle consists of four processes. First, there is a heat gain in the evaporator. Then there is a pressure rise in the compressor, followed by a heat loss in the condenser and pressure loss in the capillary. In other words, the refrigeration cycle basically consists of two heat transfer processes and two pressure change processes.

To achieve this series of processes, the heat-laden air inside the house is continuously recirculated through the coils of the evaporator. Heat from the air is transferred to the coils of the evaporator, resulting in the heat gain in the evaporator.

Next there is a pressure rise in the compressor. When the compressor is started, the pressure in the evaporator is reduced, causing the liquid refrigerant in the evaporator to vaporize or boil. Since the pumping capacity of the compressor is constant, the pressure in the low side is balanced between the inflowing liquid from the capillary and the compressor intake. This process results in the use of the heat in the air to bring about a change of state—liquid to vapor—of the refrigerant. During this change each pound of liquid refrigerant which is boiled off will absorb a considerable amount of heat, and the heat will be carried out of the evaporator with the vapor.

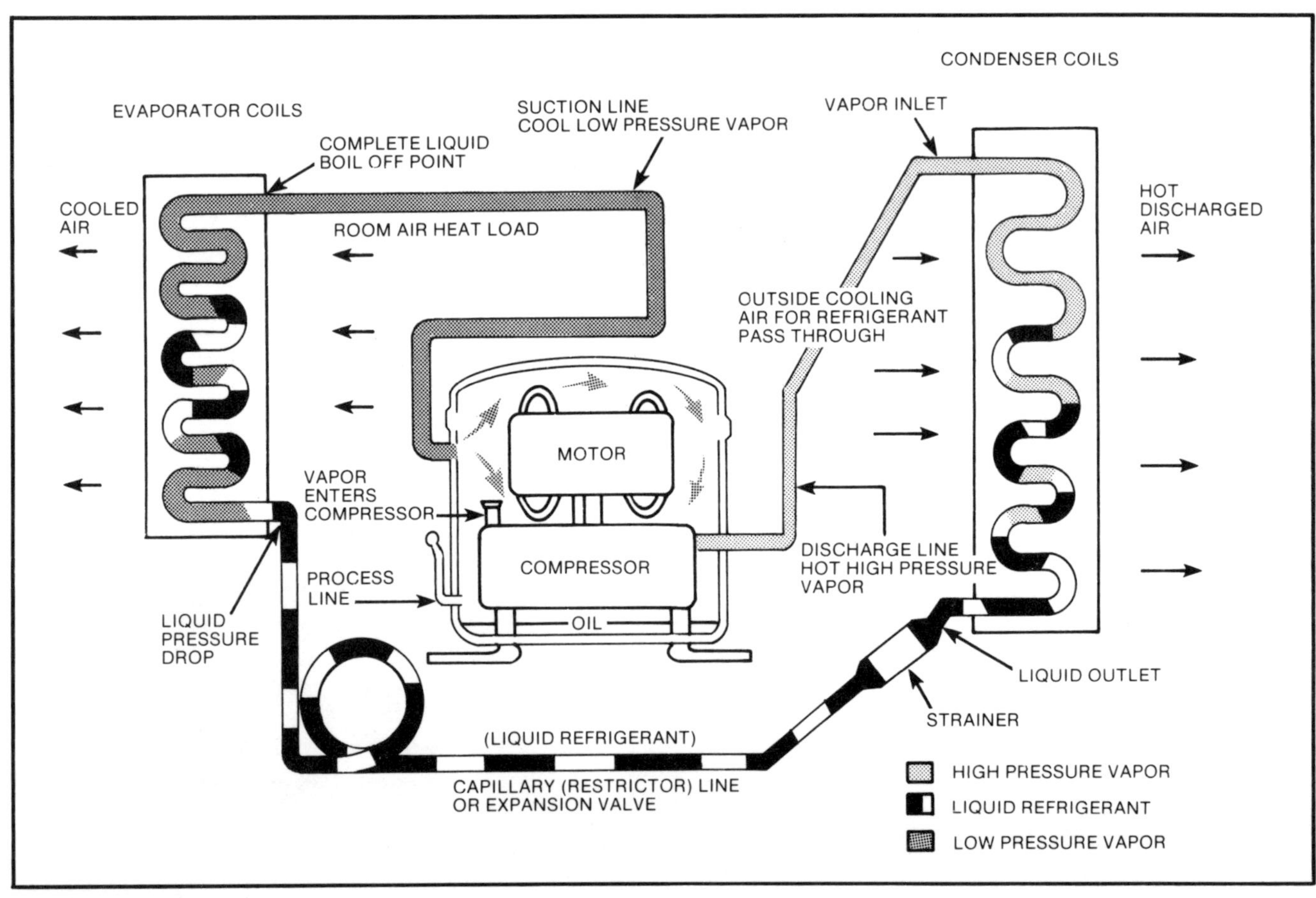

Cycle of refrigeration of a room air conditioner.

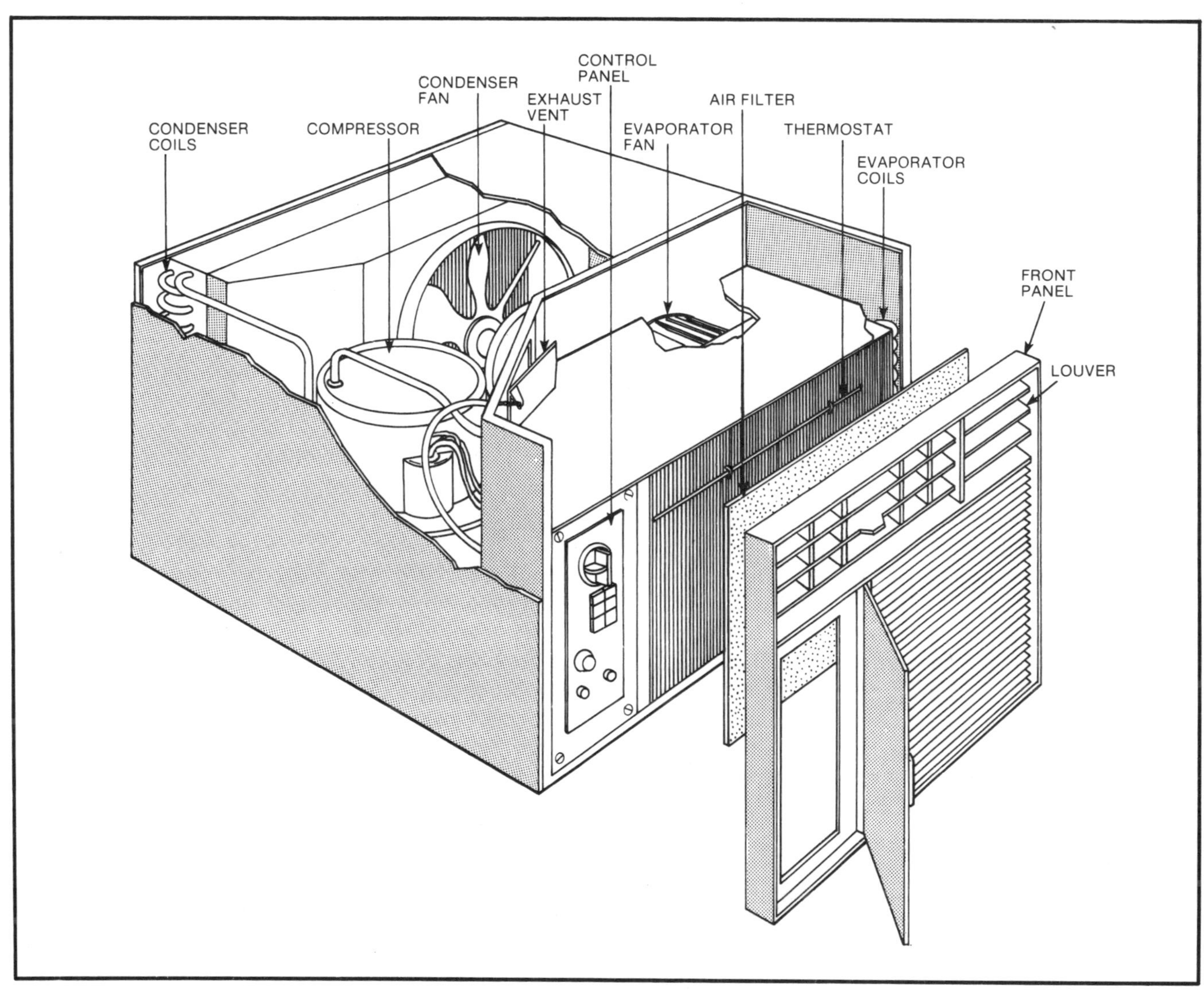

Exploded view of a typical room air conditioner.

The vapor being pumped out of the evaporator is termed a low-pressure saturated gas; it is saturated with the heat that came out of the air in the house. When the low-pressure saturated gas enters the compressor, it is compressed and forced into the condenser coil under high pressure.

The process of compression requires power from the compressor motor; this power is expended to compress the gas. After the gas leaves the compressor, it is referred to as high-pressure superheated gas.

As the hot gas passes into the condenser, heat flows from the gas to the tubing and to the fins, where the heat is transferred to the cool air flowing through the coil. Under the high-pressure condition, the heat-saturated gas undergoes another change of state, vapor to liquid. As the process of condensation takes place, the refrigerant gives up all the heat it absorbed in the evaporator and also in the compressor. Heat flows from a warm to a cool body; therefore, in order to handle the heat being removed from the condenser coil, the coil is situated in the rear end of the unit and is isolated from the air inside the house by the weatherproof bulkhead. The fan circulates outside air through the coil, and the heat is transferred to the cooler air and is carried away from the building.

The active part of the refrigeration system is the compressor. Two basic types of compressors are used in room air conditioners: rotary and piston. The piston type is most popular and is often called a low-side pump. The area inside the shell around the motor and pump is under low-side pressure. The pump pulls from the interior of the shell. High-pressure gas is discharged directly to the condenser. Some makes of room air conditioners are equipped with a rotary compressor much like an oversized refrigerator compressor. The interior of the dome is in the high-pressure side of the system. Compressor motors are generally permanent split capacitor types. They operate in the same manner as the permanent split capacitor fan motor, but are of higher horsepower and use capacitors with a higher rating. Also, the run capacitor, since it is in the circuit continuously, is usually filled with an oil which acts as a coolant.

Another primary part of the refrigeration system is the evaporator. The evaporator is a finned coil which not only cools the air, but also removes its moisture before returning it to the room. Room air is circulated through the fins and over the coils, and water vapor in this air condenses out onto the cold surfaces as a liquid and drips down into a drain pan under the evaporator. The collecting water must then be removed from the room side to the outside of the conditioned area. To do this, a room air conditioner must slope down toward the condenser; a ¼-inch drop is enough. The condensate will flow to the outside by gravity.

Room Air Conditioner Controls The controls for the room air conditioner are usually isolated from other components in a control box on the front of the chassis. The thermostat controls the compressor. It senses the temperature of the room air being drawn into the evaporator. Temperature selection is made by turning the dial knob toward a warmer or cooler setting. The user finds the point which satisfies his or her needs and adjusts it later as those needs change. Most controls use the words WARMER, COOLER, and NORMAL as reference points, but some makers use numbers to indicate temperature settings.

The majority of cooling thermostats are equipped with a vapor-filled temperature-sensing tube. This capillary-sized tube is placed in front of the evaporator and senses the temperature of the room air being drawn in for cooling. The vapor-filled sensing tube is affected by the coldest point along its entire length including the bellows. Whichever point is coldest will take over control. The thermostat body is, therefore, located in the control box away from the cold airstream. The thermostat sensing tube must not touch the evaporator or the cold-air outlet.

A second room air conditioner control is the anti-ice control. The design temperature of a room air conditioner is 95 degrees F. Operation in a room between 65 and 70 degrees F will cause evaporator icing. There just is not enough work for the air conditioner to do. The coil becomes too efficient, drops to a freezing temperature and ices up. But icing of up to 70 percent of the coil is considered normal on all models under low ambient conditions with a near-maximum cooling thermostat setting. While excess icing of the coil is usually caused by using the air conditioner when it is too cool to be effective, it could also be caused by a dirty filter, a dirty coil, or an inoperative fan. Check out these possibilities before testing the anti-ice or de-ice control for a malfunction. The purpose of this control is to prevent too much ice from building up. Generally, the anti-ice or de-ice control feeler tube is located at the center top edge of the evaporator.

The selector switch is the ON-OFF switch of the room air conditioner. It also controls fan speeds and the compressor operation. It may be a rotary or a push-button switch, depending on the particular model, but both types of switches perform the same functions. The ventilation or air-control knob is usually marked FRESH AIR, EXHAUST, or AIR CHANGER. It operates wires or chains to open and

Room Air Conditioner Troubleshooting Chart

Trouble	Probable Cause	Remedy
Air conditioner does not work.	No power at unit.	Check fuse or circuit breaker. Check for current at wall receptacle. Check line cord connection at receptacle and at switch of air conditioner.
	Selector switch is defective.	Turn switch to COOL and check for current at load side of switch. If there is current going into switch but no continuity through it, replace switch.
	Wiring within unit is defective.	Check all wiring and connections for continuity. Check switches, all thermostats, relays, overload protector, compressor, fan motor, capacitors, transformer, and so on. If any are defective, replace. Tighten any loose connections.

Room Air Conditioner Troubleshooting Chart (Continued)

Trouble	Probable Cause	Remedy
Fan runs but compressor does not and air is not cooled.	Thermostat is defective.	Check thermostat for continuity or short; replace if defective.
	Thermostat set too low.	Adjust control to proper setting.
	Loose or defective wiring or connections.	Check wiring and connections; tighten connections and replace defective parts.
	Compressor relay is defective.	Check for continuity; replace if defective.
Compressor runs and cools air, but fan does not run.	Wiring in fan circuit loose or defective.	Tighten connections or replace defective wiring.
	Fan capacitor is defective.	Check fan capacitor; replace if defective.
	Fan motor is defective.	Check moving parts in motor. Check continuity of motor windings. Replace motor if windings are open or grounded.
	Fans are obstructed or binding on shrouds or baseplate.	Inspect both fans for clearance. Adjust fan on shaft or fan motor on fan motor mounts. Check for dirt buildup on fins or bent fins; clean or straighten fins.
	Bearings on fan motor dry or seized.	Oil fan bearings; replace fan motor if bearings are worn.
Compressor stops and starts much too often; running time too short.	Thermostat bulb out of position.	Check for proper location of sensing tube or bulb in evaporator intake airstream. Tube or bulb must be in return airflow, but must not touch evaporator.
	Dirty condenser fins or restricted airflow over condenser.	Check and clean condenser; use vacuum cleaner. Check for adequate supply of fresh intake air to condenser.
	Attempting to start unit too soon after shutoff.	Wait 5 minutes before restarting to allow for pressure equalization.
	Thermostat set at too warm a temperature.	Set thermostat at a cooler temperature.
	Thermostat or anti-ice control defective.	Check temperature thermostat and anti-ice control for continuity; replace if defective.
Air conditioner does not cool sufficiently; compressor and fan motors both run.	Thermostat is defective.	Check thermostat for continuity; replace if defective.
	Thermostat set at too warm a temperature.	Set the thermostat at a cooler temperature.
	Air filter is dirty.	Clean or replace air filter.
	Restricted air to condenser.	Clean condenser and evaporator fins. Straighten any bent ones.
	Air conditioner too small for the area to be cooled.	Check that room openings are closed. If so, install unit with a higher BTU rating.

Room Air Conditioner Troubleshooting Chart (Continued)

Trouble	Probable Cause	Remedy
Air conditioner does not cool sufficiently; compressor and fan motors both run.	Refrigerating system is defective.	Low or lost refrigerant charge. Purge and recharge unit. Also, check capillary tube or filter screen for clogging or buildup of dirt.
	Cool air escaping; improper seals; doors or windows open excessively.	Readjust or replace seals. Minimize window and door openings. Close unit exhaust or ventilation doors. Close heating system registers. Seal leaks around windows and doors.
	Excessive frost or ice buildup on evaporator.	Do not operate unit when the outside temperature is below 70 degrees F.
Noisy operation.	Refrigerant tubing is vibrating or hitting adjacent tubing or metal surfaces.	Carefully align tubing slightly to eliminate vibrations and so it does not hit metal parts. Tape tubing to stop noise if necessary.
	Air conditioner insecurely mounted in window.	Check that unit is straight and that mounting brackets and other mounting parts are secured in place. Add additional supports if necessary.
	Loose, bent, or dirty fan blades.	Tighten, straighten, and clean fan blades; replace if necessary.
	Fan motor out of alignment or loose on mounting.	Check alignment. Check setscrews in fan hubs. Check fan motor mounting. Tighten if loose and replace grommets if worn.
	Fan motor bearings are dry or defective.	Lubricate fan motor; replace motor if still excessively noisy.
Evaporator frosts or ices over.	Restricted airflow over evaporator.	Clean air filter, air passages, both fan wheels, and evaporator coil. Straighten evaporator fins if bent.
	Outside temperature below 70 degrees F.	Do not operate unit when the outside temperature is below 70 degrees F.
	Undercharge of refrigerant.	Purge and recharge unit.
Water drips from air conditioner.	Cabinet not tilted properly.	Cabinet should tilt at least ¼ inch toward outside.
	Condensate drain from evaporator to condenser area is plugged.	Clean condensate port with a small brush.
	Extreme humidity.	Dripping is unavoidable; attach a pan and drain to carry water out of the way if necessary. Minimize door openings.
	Inadequate seal.	Check and improve all sealed areas including gaskets around window duct.

Room Air Conditioner Troubleshooting Chart (Continued)

Trouble	Probable Cause	Remedy
Air conditioner repeatedly blows fuse or trips circuit breaker.	Incorrect size fuse.	Use time-delay fuse of the correct rating.
	Air conditioner circuit overloaded.	Plug other appliances into another circuit.
	Short circuit.	Check line cord, wiring, capacitor, and compressors for short circuit.
	Turning air conditioner on too soon after turning it off.	Wait five minutes before restarting to allow for pressure equalization.

close dampers in the bulkhead between the evaporator and the condensing unit and determines the mix of cooled room air and warmer outside air that will be circulated in the room.

Preventive Maintenance Filters are an important part of an air conditioning unit. They provide that the air being blown into the room is always clean and they prevent dirt from building up in the air conditioner. The most common loss of cooling and ventilating efficiency in an air conditioner is caused by a dirty or clogged filter. Filters must be inspected periodically and thoroughly cleaned when dirty. To inspect a filter, hold it up to the light. If the light is not clearly visible through the filter mesh, clean the filter. In cleaning extremely dirty filters, the bulk of the dirt may be removed with a vacuum cleaner. Filters should then be held under a stream of warm water, clean side up, to remove the dirt embedded in the filter mesh. However, bleach solutions or dry detergents should never be used to clean the filters. When replacing filters, be sure the thermostat sensing tube is in its proper position. Never operate an air conditioner without a filter. Dirt will enter the evaporator coil, clogging it and reducing cooling capacity.

Another common problem you will probably encounter is fins being bent if someone bumps into the air conditioner. Bent or dirty condenser or evaporator fins cut airflow, reducing cooling efficiency. Straighten any bent fins with a fin comb, which you can buy at refrigeration parts stores. The spacing of the comb teeth must match that of the fins on the coils of your unit. These combs also remove dirt from the coils, but a vacuum cleaner does a more thorough job.

Washing Machines

Along with refrigerators and ovens, the washing machine is one of the appliances we have become most accustomed to using and most dependent on. When your washing machine breaks down, your options are to wash your laundry by hand or take it out to be washed—both relatively undesirable choices. Although a washing machine is not a simple piece of machinery, your return for fixing even minor malfunctions will be great. The cost of having a professional serviceman come to your house and make even a minor repair on a washing machine is almost in every case so exorbitant as to be prohibitive. For example, you can replace a component such as a timer fairly inexpensively and the job should not take a great deal of time. However, if you were to have a repairman come to your

A typical automatic top-loading washing machine.

house, you would be charged for his diagnosis and repair time, plus the cost of the timer, which includes a big markup. If he did not have the part in his service truck, he would have to make a second trip to your house and you would be charged for that as well. Washing machines, like refrigerators, ovens, and dishwashers, may seem hard to repair; but if you make the effort, your savings could be very substantial.

Operation The soil/dirt removal operation in a modern washing machine is a combination of chemical and mechanical processes. First the chemical action of the detergent or soap solution dissolves and loosens the soil in a fabric. The mechanical action involves flexing the clothes and forcing the detergent or soap through the fabric, which in turn removes the soil. The operation of the washer is aided by heat and by the softness of the water, which increases the chemical action of the detergent or soap. Almost all modern automatic washers employ one of two types of mechanical action: tumbler, also called cylinder action, or agitator action. The latter is by far the more popular and more commonly used today and will be the type we will discuss. But all automatic washers, regardless of type, model, or make, have only four basic operations: fill, wash and rinse, pumpout, and extraction.

The heart of the agitator-type washing machine is the agitator, which usually consists of vanes or blades on a cone that fits over a central shaft in the washer tub. As the agitator turns back and forth, the blades or vanes catch the clothes and move them about. This movement also creates currents in the water, which contribute to the cleaning action. There are a great variety of agitator designs; they have vanes or blades of various numbers, designs, and sizes, which are arranged in a vertical or spiral position. Agitators may be made of solid or perforated plastic or a metal which is usually aluminum.

Most agitator-type washing machines employ an oscillating—back-and-forth—action during the wash cycle. Generally, to produce this oscillating action, an arm, usually called a rack bar or Pitman arm, is connected off-center to a low-speed gear wheel. As this gear wheel turns, it imparts a back-and-forth motion to the arm. This motion, in turn, is transmitted to a pinion gear which drives the agitator. There are also other methods of driving the agitator. For instance, a few models provide a slow-speed, off-center, wobbling motion to the agitator, while some others impart an up-and-down, pulsating motion to it. While the oscillating action is the one most commonly used for the washing operation, most machines of this type employ a rotating or revolving motion to spin the tub or basket for the extraction operation. To accomplish this, a clutch action of some type is used to disengage one set of gears and engage the other. Incidentally, agitator-type washing machines are always top loading, meaning that the clothes are placed in the washer through a door or lid that opens on the top of the unit.

The needs and components of most washer models are about the same. They all require hot and cold water, which is fed into valves in the washer that turn on and off the hot and cold water and mix them at appropriate times. While a few washers control water temperature with a thermostat, most operate on a simple on-off principle. That is, when the hot water is on and cold is off, the water in the washer is hot—whatever temperature the water heater tank provides, which is usually about 150 degrees F. When the cold water is on and hot is off, the water in the washer is cold—whatever temperature the cold water tap provides. When both hot and cold are on, they are evenly mixed to provide warm water. With average cold water temperatures out of the tap about 50 degrees F, the mixture generally comes out somewhere around 100 degrees F.

All automatic washers have an electric motor as well as a pump. The motor on most models operates in both counterclockwise and clockwise directions when viewed from the top of the machine. It operates counterclockwise during the wash and rinse cycles and clockwise during the pumpout and spin operations. The motor turns the pump and drive pulleys through a belt arrangement. At the end of the pumpout period, a solenoid releases the clutch spring and the spin basket rotates to extract the water from the clothes. When the agitator is in operation, power is transferred directly into the transmission from the drive pulley through the transmission drive shaft and the clutch spring located inside the transmission case. During pumpout and spin periods the clockwise rotation of the motor releases this clutch.

Solenoids play a very important part in the operation of an automatic washer. In addition to operating the clutch and gearshift arrangements, they control water flow, detergent applications, and the like. For example, mixing valves fill the washing machine with hot, cold, or warm water, depending on which valve has been opened. When a valve is closed, a solenoid plunger closes a pilot hole in a flexible diaphragm and water from a valve port enters an area above the diaphragm through a bleed hole. The water pressure is the same on both sides of the diaphragm, but the overall pressure on top is greater because the pressure there includes a greater area; this keeps the valve closed. To open a valve, the solenoid raises the plunger to open the pilot hole in the diaphragm. Water flows from the area above the diaphragm, lowering the pressure there. Water below the diaphragm now can lift it

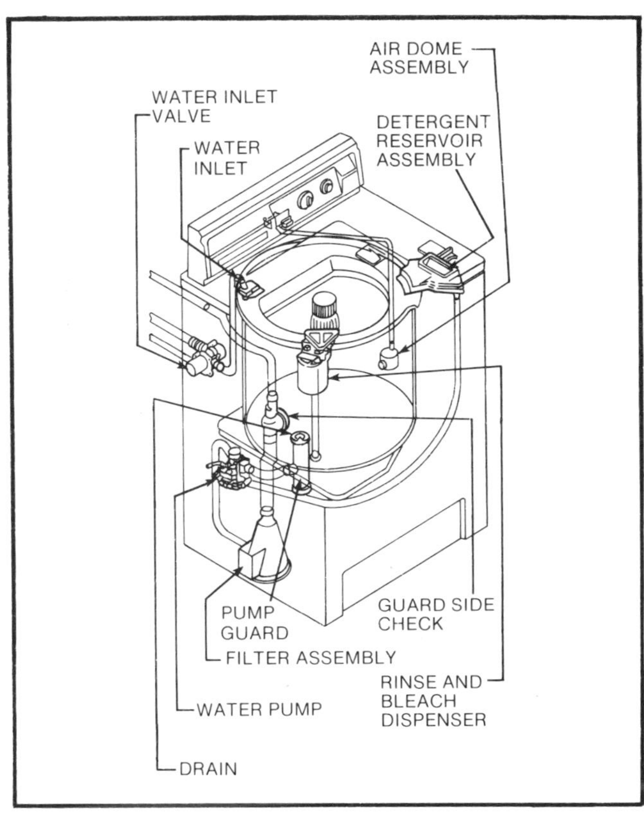

Cutaway view of a top-loading washing machine.

and flow into the tub. Of course, the overall control of the automatic washer is left to the timer. While part of the control is selected by the user—for instance, washing time and water temperature—most of the automatic action is performed at certain preselected time intervals by the timer.

Repairs and Servicing Because a washing machine is a complicated machine, the first step in the repair process is to isolate the malfunctioning component. Despite the washer's complexity, it is like all other large appliances—a collection of simpler parts, most of which you will be able to deal with individually. If the tub will not fill up with water, a possible cause is that the filters at the hose couplings are clogged with hard-water minerals or debris. These filters should be cleaned occasionally. Another possibility is that the inlet valve screens are dirty, in which case the debris and slime should be removed with a stiff-bristle toothbrush and running water. If the filters are clean, a solenoid plunger may be stuck or otherwise defective. The solenoids can be tested electrically by checking for continuity across the two terminals. Other possible parts that may be malfunctioning if the tub does not fill with water are the fill, float, or water level switch or the water temperature switch. These switches can also be checked for continuity with your VOM if you suspect a problem.

If you have power to your washing machine, but it does not run, try different combinations of cycles and water temperatures to see if the machine responds. In many cases it will. If, for example, you had selected the warm water permanent press cycle, there is a chance that the water mixing solenoid at the inlet valve is open or that a fill switch is defective. The last component to suspect is the motor. The capacitor-start motors found in washing machines are generally extremely dependable and require little attention.

A fairly common failing is a broken or loose drive belt; however, broken belts are easy to spot and loose belts are easy to diagnose. A properly adjusted belt should have about ½ inch of play halfway between the pulleys. When you depress the belt at that point, it should move through the ½ inch without too much pressure. If you find it loose, loosen the adjuster lock nut and move the motor away from the opposing pulley. Check the tension after tightening the nut. If your adjustment did not tighten the belt to the prescribed tension, check to see that the idler pulley is functioning. Some washing machines use a spring-loaded idler to help maintain the correct tension. If your machine does not have one of these and the belt will not tighten after your adjustment, you will have to install a new belt.

The timer is the brain of a washing machine. If any part of the timer fails, you can be assured that you are going to have problems with the machine. In the operation of the timer, a small synchronous motor advances cam wheels through a gear train, called an escapement, in sudden movements every minute or two. The less than fluid motion reduces contact point arcing by breaking contact quickly instead of slowly. When the timer refuses to advance, try putting a single drop of light motor oil on the timer motor shaft and on the escapement shafts. The timer is normally mounted to the inside of the control plate. If you remove it and find an open or grounded timer motor, burned switches, or a broken escapement, you will have to replace the entire timer unit.

Most water leaks originate from hose joints, but you might have to do a little exploring because a small leak from a hose joint can run along the hose for some distance before it drips off. If you open the cabinet and watch the machine during its wash cycle, you stand a good chance of spotting any water leaks. After some years your washer may also develop a leak or two in the tub bolt holes, where it attaches to the chassis. Tubs are porcelain coated, and the weakest part of the coating is at these bolt holes. The porcelain might have been chipped during assembly, allowing rust to grow and a leak to develop. Most manufacturers sell repair kits consisting of large-headed brass bolts and neoprene washers to fix this type of leak.

An open or binding gear-change solenoid, an open motor-reversal circuit, and a broken belt are three probable causes when your washing machine will not spin. If you remove the rear inspection panel you will spot a broken belt immediately. Otherwise, check the solenoid that engages the spinning clutch or check the spin cycle circuit for continuity, beginning with the timer switch contacts. To be sure that the timer is advancing, run the machine through a complete washing program. When it reaches the spin stage and nothing happens, try advancing the timer by hand. If the tub begins to spin, then you know that the timer is binding at this point.

Excessive noise and vibration can be caused by a wide variety of problems, but two common ones are an imbalanced load in the washer and a washer that is not situated on a flat surface. The solutions are simple. Be sure to load your washer according to the manufacturer's instructions for the best balance and adjust the leveling feet with the help of a carpenter's level. Check the level from side to side and front to back. If your washer does not have leveling feet, use shims to level it.

Washing Machine Troubleshooting Chart

Trouble	Probable Cause	Remedy
Washer does not work.	No power to washer.	Check to see if fuse is blown or circuit breaker is tripped. Correct circuit overload. Make sure washer is plugged in; if it is, check line cord for continuity.
	Out of balance or safety switch activated.	Be sure clothes are evenly distributed in the tub and the washer door is closed.
	Selector switch is defective.	Check selector switch; replace as needed.
	Timer is malfunctioning.	Check timer; repair or replace if defective.
	Motor overload protector is defective.	Replace protector assembly in motor; in built-in types, the entire motor must be replaced.
	Pressure or water level switch is defective.	Check switch; replace if defective.
Motor has power but will not turn in either direction.	Too large a load.	Wash smaller loads.
	Too many suds.	Use less detergent or a low-sudsing one.
	Tight pump impeller because of foreign object in pump or pump is seized.	Must remove restriction to free impeller or replace pump.
	Motor speed selector switch inoperative or leads broken.	Check switch; repair or replace as needed.
	Motor is defective.	Check motor for continuity or short; repair or replace.
	Bearing or bushing in drive mechanism broken, too tight, or not lubricated.	Replace parts and/or lubricate as needed.
	Timer is defective.	Check timer; replace if defective.
Motor has power but will not spin the basket.	Belt broken or off a pulley.	Replace belt.
	Belt is loose.	Adjust motor bracket to tighten belt; if this does not solve problem, replace belt.

Washing Machine Troubleshooting Chart (Continued)

Trouble	Probable Cause	Remedy
Motor has power but will not spin the basket.	Load unevenly distributed in the tub.	Reposition load.
	Timer is defective.	Check timer; replace if defective.
	Safety spin switch is out of adjustment or defective.	Adjust switch bracket or lever or change switch as necessary.
	Clutch loose, clutch spring broken, clutch spring housing bent or worn, or clutch brake is broken.	Check; replace defective parts or tighten clutch.
	Spin solenoid is defective.	Check; replace solenoid.
	Broken gear in transmission.	Gear must be replaced.
Washer will not agitate.	Loose or broken drive belt.	Tighten or repair belt.
	Motor is defective.	Check motor; replace if defective.
	Agitator solenoid is defective.	Check; replace solenoid if defective.
	Loose drive or motor pulleys.	Tighten pulleys.
	Timer is defective.	Check timer; replace if defective.
Water does not enter the tub.	Washer not actually turned on.	Rotate the timer dial a bit farther, or push selector button more firmly.
	Water supply turned off.	Open valves or faucets.
	Supply or inlet hoses kinked or otherwise obstructed.	Straighten, remove obstruction, or replace hose.
	Strainers in mixing valve are clogged.	Remove and clean or replace strainers.
	Water level switch is defective.	Check for continuity; replace if defective.
	Inlet valve solenoid is defective or lead is loose.	Secure terminal or replace solenoid on valve.
	Dirt particles in mixing valve.	Tap valve or switch it on and off in FILL position to dislodge dirt.
	Float switch is stuck open.	Repair or replace float switch.
	Valve diaphragms worn or damaged.	Replace diaphragms in valve.
	Timer is defective.	Check; replace timer if necessary.
Washer leaks water.	Inlet hose not secured to inlet valve.	Tighten hose connections.
	Drain connection or drain not secure to pump.	Check hose clamp; repair or replace hose connections as necessary.
	Hose is defective.	Replace defective hose.
	Pump gasket not in place.	Replace gasket.
	Tub overfilled because of defective timer or pressure switch.	Change, adjust, or repair defective parts.
	Tub leaks.	Repair or replace tub.

Washing Machine Troubleshooting Chart (Continued)

Trouble	Probable Cause	Remedy
Wash or rinse water is not hot enough.	Hot water supply is exhausted.	Check water heater thermostat; make sure it is set at 140 to 160 degrees F. Do not attempt to use more hot water than your system can supply.
Water does not drain out of washer or drains very slowly.	Suds lock.	Dip excess suds out of washer and wash out with cold water to remove suds.
	Filter plugged.	Replace filter.
	Water pump dirty or defective.	Clean pump or replace it.
	Drain solenoid is defective.	Replace drain solenoid.
	Drain hose kinked or clogged.	Straighten and relocate hose to prevent kinking; remove and flush out if clogged.
	Pump belt broken or loose.	Adjust or replace belt.
	Drain hose too high.	Lower position of drain hose.
	Timer is defective.	Check timer; replace if necessary.
Washer vibrates excessively or walks across the floor during spin operation.	Washer not level.	Adjust leveling legs from side to side and front to rear; use shims or wooden blocks if necessary.
	Load out of balance.	Stop washer and redistribute load; always distribute clothing evenly around agitator.
Washer damages fabrics or tears clothes.	Washer overloaded.	Wash smaller loads.
	Water level too low.	Increase water level.
	Use of excessive bleach.	Use less bleach and never pour undiluted liquid bleach on clothes.
	Clothes put in washer which should be washed by hand.	Determine which clothes should be hand-washed.
	Using agitation longer than required or at too high a speed for fine fabrics.	Use correct settings for fine or delicate fabrics.
	Clothes catch under agitator on some models.	Agitator too loose or set too high; adjust as needed.
Washer does a poor job of washing.	Improper water extraction.	Avoid out of balance loads; also, a slow spin may be caused by a faulty clutch, tight transmission, or kinked drain hose. Check and take necessary action.
	Oversudsing.	Use correct type and amount of detergent.
	Insufficient detergent for dirt in clothes or for the hardness of water at your home.	Use correct measured amount of detergent for each load; as the hardness of water increases, the amount of detergent needs to be increased.

Washing Machine Troubleshooting Chart (Continued)

Trouble	Probable Cause	Remedy
Clothes still wet after final spin.	Clutch slipping.	Check and repair as needed.
	Load out of balance in tub causing clutch to slip; tub does not reach full spin.	Redistribute load and run through spin cycle again.
	Drain hose restricted.	Remove obstruction.
	Low spin selected which is not suitable for load.	Reselect spin cycle.
Washer blows fuse or trips circuit breaker.	Overload during agitation.	Do not put too many clothes in washer at one time or use a water level that is too low.
	Short circuit.	Check for continuity until short circuit is found; replace defective part.
	Clutch stuck.	Correct as needed.
	Timer is defective.	Replace timer.

Clothes Dryers

All clothes dryers are automatic and have control centers with buttons or dials that allow you to program the unit to handle different types of loads. For example, the selector switch might have separate cycles for drying cottons, delicate drying, or drying permanent press fabrics. The drying temperature and time of drying can also be selected according to the amount and kind of clothing you plan to dry. Dryers now have lint filters, while some also have a two-speed selection for light and heavy clothing as well as cool-down phases to reduce wrinkling. A clothes dryer consists of a motor-driven revolving basket, an electric heating element, thermostats, and a timer. Some models also have a selector switch.

Operation Air, heated by the electric heating element, is forced through the tumbling clothes by a fan; the fan in turn is driven by the drive motor. The temperature of the heated air, entering and leaving the basket, is controlled by thermostats which maintain a balance between the air velocity, air volume, and air temperature. The temperature of the exhaust air is a measure of the dryness of the clothes.

After the dryer is loaded with damp articles and started, the temperature inside the basket will rise rapidly. When the temperature reaches approximately 130 degrees F, evaporation of the moisture in the load will absorb the heat as rapidly as the heat is generated by the heater. The temperature will not rise appreciably above 130 degrees F until the load is nearly dry. When the load is nearly dry, there will not be enough moisture in the clothes to absorb the heat, and the basket and clothing temperatures will rise. This heat rise will continue until the exhaust air reaches approximately 160 degrees F. At this point the heater circuit is disconnected by a thermostat. The dryer basket will continue to

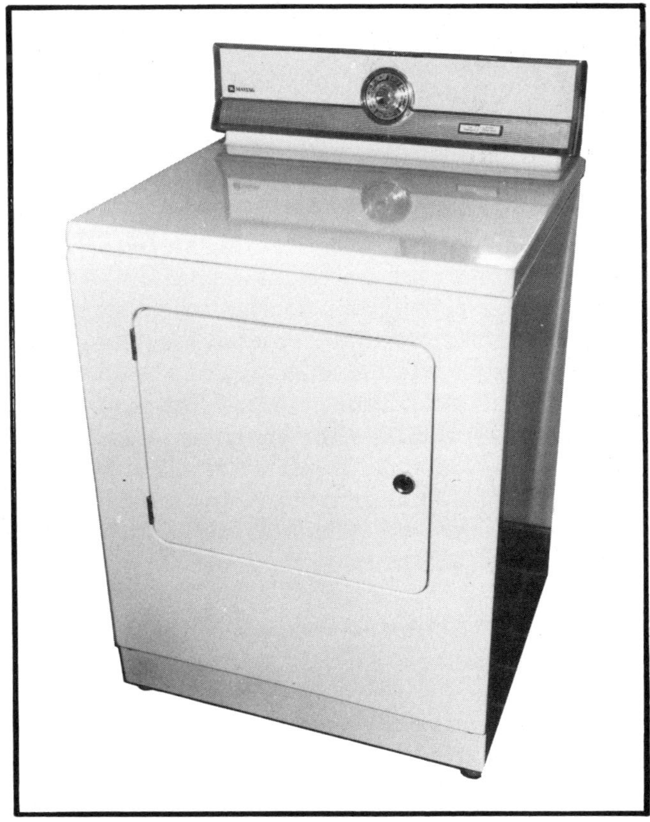

A typical automatic clothes dryer.

revolve in what is called the overrun period, and cool, fresh air will enter the basket to cool the clothes for handling. The overrun time may last from 3 to 10 minutes and is controlled by either the timer or an overrun thermostat which will stop the dryer.

There are two basic types of timer drying operations: time dry and automatic dry (usually called auto-dry). In the time dry method, you select a particular length of time to dry the clothes. The timer motor and the heat source operate at the same time. The dryer continues to operate for that length of time, even after the clothes are dried. Auto-dry eliminates unnecessary drying time because the operation of the auto-dry cycle is controlled by the cycling thermostat, a single-pole, double-throw switch, connected to the heat source on one side and to the timer motor on the other side. When the circuit to the heat source is closed, the timer motor circuit is open. When the heater circuit opens, the circuit to the timer motor closes, and the timer advances to the OFF position. When the clothes are dried, the circuit in the timer to the heat source opens, and a second circuit directly to the timer motor closes and advances the timer to OFF. However, for proper and effective use of the auto-dry cycle, the location of the dryer, if it has this feature, should not be in an unheated place where the dissipation of the drum heat will not allow the cycling thermostat to cycle properly and, therefore, not allow dryer timer to advance to the OFF position.

Repairs and Servicing Many times a malfunctioning dryer is caused by simple trouble spots. For example, one of the largest contributors to inefficient drying or dryer failure is lint. All dryers are designed to pass a predetermined volume of air through the clothes during the drying cycle. This air is controlled by thermostats. Any restriction in the system, such as lint, will throw the air-heat volume out of balance. This will result in such problems as long drying time, damp clothes, burned-out elements, thermostat failure, or a burned odor caused by lint particles that catch in the heater ducts and scorch. To deal with the potential lint problem, all dryers are equipped with lint traps or collectors, usually a fine mesh bag with an elastic opening or some type of fine filter mounted in a housing built into the discharge air stream. It is extremely important to keep the filter and the dryer clean of a lint buildup.

The final electrical component in an automatic dryer and perhaps the heart of the operation is the timer. Each dryer timer consists of two basic components assembled into one unit. The components are the motor and escapement, and the switch box. The escapement houses the speed reduction gears, and the switch box contains the cam and switch contacts. Timer motor escapements may vary slightly because of different sources of manufacture, but all function in the same manner. The speed reduction gears in the escapement reduce the timer motor speed to about one revolution per hour for most dryers. Although this can vary, sometimes a timer can malfunction but the dryer will still operate. For example, if your dryer starts and operates but does not shut off in conjunction with the timer setting, the timer is defective and must be replaced. This can usually be accomplished by removing the entire timer unit from the front of the dryer and simply replacing it with a whole new unit.

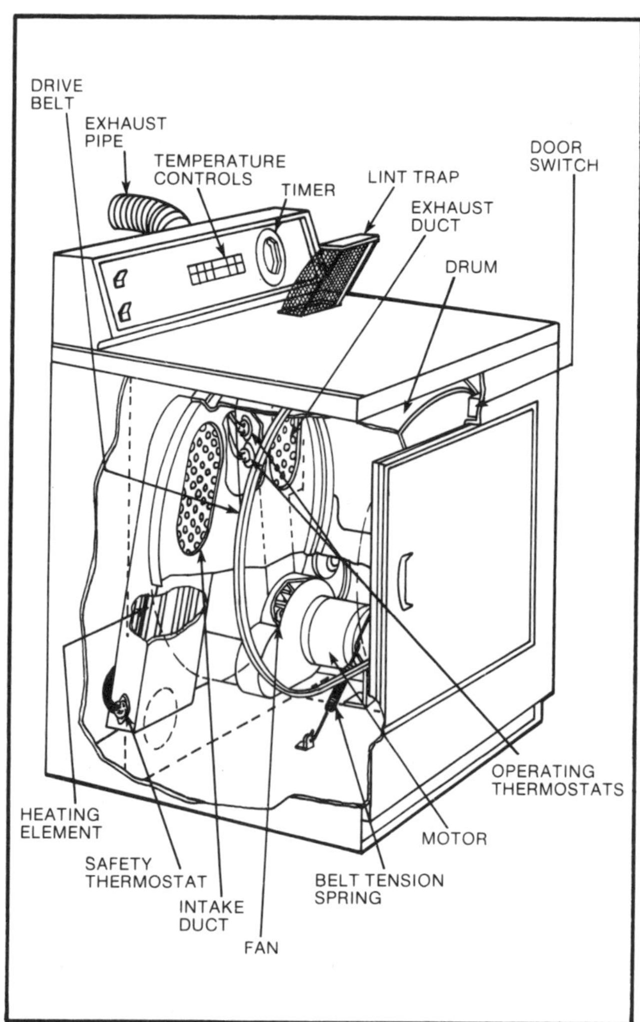

Cutaway view of an automatic clothes dryer.

Automatic Dishwashers

Anyone who has one will agree that automatic dishwashers have eliminated a great deal of drudgery in the kitchen. Today there are two basic types of automatic dishwashers: top loaders and front loaders. As their names imply, the top loaders are loaded through doors which are located on the top of the unit, while front loaders are loaded through

Clothes Dryer Troubleshooting Chart

Trouble	Probable Cause	Remedy
Dryer does not work.	No power.	Check line fuses or circuit breakers or line switches.
	Line cord is defective.	Check line cord for continuity; replace if defective.
	Door switch is defective.	Check door switch for continuity; replace if defective.
	Loading door not closing tightly or not actuating door switch.	Correct condition to ensure proper door action.
	Timer is defective.	Check timer; replace if defective.
	Start switch is defective.	Check start switch for continuity; replace if defective.
	Motor does not work.	Check and replace if faulty.
	Selector switch is defective.	Check selector switch; replace if defective.
Motor runs but drum does not rotate.	Belt broken, loose, or off drive pulleys.	Replace belt on pulleys.
	Broken tension spring.	Replace spring.
	Drum pulley loose on shaft.	Reposition pulley; tighten pulley on drum shaft.
	Motor or idler pulley loose.	Reposition pulley and tighten the shaft.
	Drum binding.	Locate cause of binding and correct.
Dryer runs with the door open.	Door switch is defective.	Replace door switch.
	Door switch lever or plunger bent or binding.	Locate cause of binding and correct.
Motor continues to run when timer is off.	Motor is grounded.	Find problem and correct.
	Timer is defective.	Check timer; replace if necessary.
	Operating thermostat is defective.	Check for defective thermostat; replace if necessary.
	Selector switch is defective.	Check selector switch; replace if faulty.
Dryer motor runs but dryer does not heat.	Operating or safety thermostat is defective.	Check thermostats for continuity; replace defective thermostat.
	Timer is defective.	Check timer; replace if defective.
	Heating element is defective.	Check heating element for broken or sagging coil or broken insulators.
	Motor centrifugal switch is defective.	Check switch for continuity across terminals; replace if defective.
Dryer runs and heats, but clothes are not properly dried.	Lint screen is clogged.	Clean lint screen and keep it cleaned.
	Exhaust duct is blocked.	Clean lint from exhaust duct.

Clothes Dryer Troubleshooting Chart (Continued)

Trouble	Probable Cause	Remedy
Dryer runs and heats, but clothes are not properly dried.	Clothes too wet when placed in dryer.	Clothes dryers are designed to dry damp, not wringing wet clothing. Heavy, wet clothes can overload the drum and motor.
	Load of clothing is too large.	Do not load dryer too full.
	Drum seals not tight.	Check front and rear drum seals.
	Door not sealing properly.	Locate cause and correct.
	Exhaust fan not functioning adequately.	Check fan; make sure it is free of lint or otherwise unobstructed.
	Operating or safety thermostat is not working.	Check thermostats; replace if defective.
	Main motor not operating properly.	Check centrifugal switch and replace if faulty.
	Timer is defective.	Check timer; replace if necessary.
	Drum belt not functioning properly.	Check belt for breaks, wear, or being off a pulley; also check for faulty idler assembly. Replace defective parts.
	Selector switch is defective.	Check switch; replace if necessary.
Dryer smokes.	Overheated motor.	Check motor; correct as necessary.
	Shorted wires.	Check for short; find cause of short and replace faulty parts.
	Excessive lint accumulation in cabinet.	Clean all lint from dryer.
Signal device does not work.	Signal device has open circuit.	Check for continuity and replace device if faulty.
	Loose connection.	Check wiring and tighten connection.
Dryer blows fuse or trips circuit breaker.	Short circuit.	Check electrical circuit for short until defective part is found and replaced.

front bottom-hinged doors which open in the same manner as oven doors. Top loaders are portable; they are freestanding and are mounted on casters. They can be wheeled near the dining area, loaded with dishes, and then moved next to the sink. Here the water connections are made, the line cord plugged into an electrical outlet, and the unit turned on. Once the washing, rinsing, and drying operations are completed, the dishwasher may be stored anywhere.

The front loaders are undercounter built-ins or convertible. The built-in or permanently installed models are standardized in size to fit under the standard kitchen countertop with base cabinets. This model is usually located next to the sink and has permanent water and electrical connections. A convertible dishwasher comes on casters, but can be installed under the counter later if the owner decides to make a permanent arrangement. In general, the principles of operation and the resultant servicing problems are similar in these types except that the portable and convertible models require electrical and water connections to be remade each time they are used. The presence of a flexible attachment service cord and hoses with these types introduces the possibility of openings in the cord and a poor connection at the outlet, not usually problems with the undercounter models.

A typical automatic dishwasher.

Operation The basic principles of operation are fairly simple. In almost all models, the washing action is accomplished by means of a strong spray of hot water coming up from the bottom. It hits the dishes from the bottom directly and from the top and sides indirectly by bouncing off the inside of the tub. There are two methods of obtaining the necessary strong spray of water. In one, a revolving spray or wash arm in the bottom with jets (perforations) from which water emerges provides the washing action. The other method, often called the splash type, involves the use of an impeller blade (vane) in the bottom which hurls the water at the dishes and cookware. In recent years, the spray arm action type of dishwasher has become much more popular and is the type we will consider.

In a typical spray-type dishwasher with a wash arm, the pump and motor assembly is located in the sump at the bottom of the dishwasher. When the pump motor rotates clockwise, the lower or drain pump impeller, rotating backwards, does not pump. The upper impeller, rotating forward, picks up water from the sump and propels it upward into the spray arm and spray column. When the motor turns counterclockwise, the upper impeller windmills and the lower impeller pumps the water and food particles in the sump out of the drain hose. The drain line loop is quite important because it prevents the water from draining out by gravity during the wash or rinse cycles and prevents the drain impeller from forcing any water out while it is windmilling during the wash or rinse cycles. An equivalent to the loop height which is about 30 to 32 inches is provided on mobile models by connecting the drain hose to the faucet coupler.

Water is admitted to the dishwasher by a timer-controlled solenoid valve which controls the cycles of the dishwasher. Detergent is usually provided automatically from a detergent dispenser. In some models, there is a rinse additive dispenser which releases a small amount of additive into the final rinse to help protect glassware from the spotting caused by minerals in the water. The rinse additive acts as a wetting agent, causing the water to run off the dishes without spotting.

Air vents, located at the bottom rear of the tub and in the door, provide for the circulation during the drying cycle. This drying is accomplished by means of condensation. In this process, the room air cools the outside of the tub in the same way that a cold winter day cools the windows of a house. The windows will sweat on such a day; the windows are wet but the warm air in the room is dry because most of the moisture in the room has condensed on the cool windows. The same thing happens in the dishwasher. Room air is cooler than the air in the dishwasher, so it cools the tub walls, and the moisture inside the dishwasher condenses on the cooler walls. Drops of water may be on the inside of the door when it is opened, but the dishes will be virtually dry.

To accomplish its task, the dishwasher performs three basic functions. The first function is the wash/rinse, in which the dishwasher fills with water and recirculates it. A typical fill admits about 6 to 9 quarts of water, flowing at a rate of about 1½ gallons per minute. Depending on the cycle, detergent or a rinse additive may be put in at a specified time. The second function is to drain. While there are two types of drainage systems in use with dishwashers, the pump drain is used a great deal more than the gravity drain system. In this function, the pump motor stops, changes direction of rotation, and pumps the water out of the dishwasher. The third function is to dry. The heater may be on all the time or may cycle on and off to provide a lower temperature, depending on the cycle selected.

Repairs and Servicing A variety of problems associated with an automatic dishwasher are capable of cropping up. The usual contact problems will occasionally cause the door switch or cycle-selector switch to malfunction. They can be checked with the usual continuity tests. The water fill valve can suffer both electrical and mechanical failures. Mechanical problems can usually be traced to dirt in the line or corrosion buildup. When mineral deposits build up on the valve armature, spring, armature guide, and diaphragm, it is best to replace these parts rather than to try to clean them. The timer is the brain of the dishwasher. It programs the sequence in the dishwasher, accomplishing the washing and drying cycles. However, the timer is also an Achilles' heel because a number of things can go wrong with it and if it malfunctions, the dishwasher will probably not work very well. If you suspect the timer is defective, make a visual check of the contacts. Then manually check that the contacts come together and break apart properly. Also check the condition of the cam teeth. Put power to the timer motor to see if the field rotor and gear train are operative. If they are not operating, check the field for continuity and for the possibility of bind between the gear train and clutch. If the timer fails any of these checks, it must be replaced.

The heating element will also occasionally become defective. To verify that the heating element is working, allow the dishwasher to cycle until it enters the drying cycle. Open the door after a few minutes and sprinkle some water onto the heating element. There should be a noticeable hiss as the water comes in contact with the element. If not, check the heating coil for continuity and wattage. Another possible failure is in the detergent dispenser. However, if the lid assembly is sticking occasionally, you should check to see if detergent is sticking in the latching mechanism, in which case it should be brushed out. Also, putting a small amount of silicone lubricant on the shaft and the spring catches of the lid assemblies is a good form of preventive maintenance.

To check a typical pump assembly during the recirculation portion of its action, first inspect the sump inlet strainers, drain both and recirculate. Clean if required. Allow the washer to fill and check the water level. Check the spray arm rotation by opening and closing the door several times during a wash or rinse period and noting the position of the spray arm each time to establish whether the arm is rotating. Rotation should be a minimum of 40 rotations a minute for the main spray arm. Where auxiliary spray arms are used, their mere rotation is all that is required for effective wash action; no specific arm speed is required. To further test for circulation with empty dishracks and a normal water fill, place an empty tea cup in each corner of the upper rack in such a position as to retain the recirculating water. Operate the dishwasher in WASH or RINSE for approximately 20 to 25 seconds and note whether the cups are filled. This test, although not completely accurate in checking pump

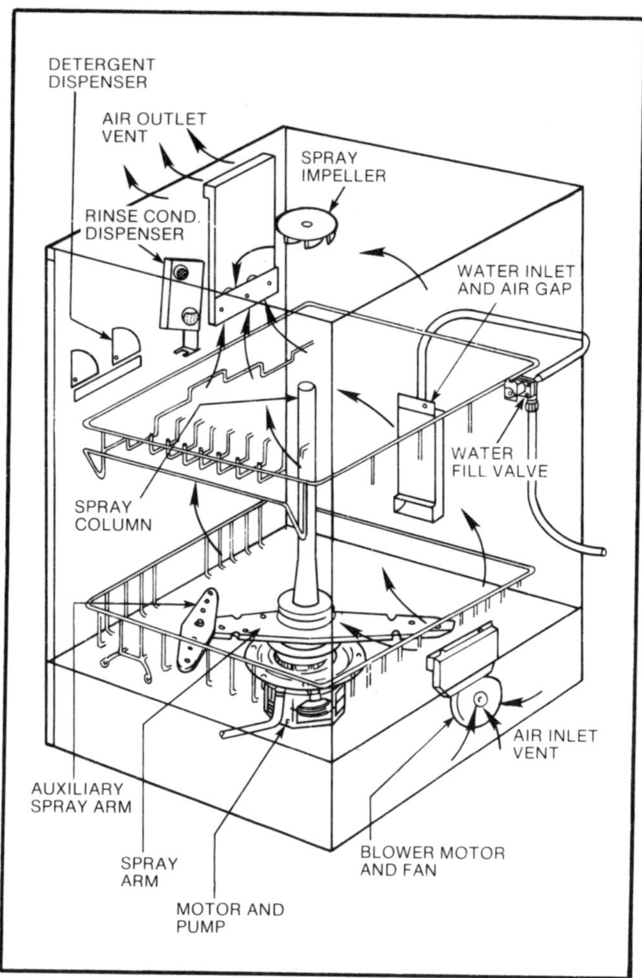

Major parts of an automatic dishwasher.

capacity, can determine whether water is reaching the critical corner areas.

If a problem of circulation is indicated, check for a sticking spray arm, interference between the spray tower and dishrack, or a restricted pump inlet retainer. Many models of dishwashers have only one washer under the spray arm retaining nut. Several additional washers may be added, if necessary, to keep the spray arm from hitting the heater. You should also check the relay coil in the starting relay for continuity if your dishwasher will not start or operates on thermal overload when trying to start.

Filter screens are normally located over the pump entrance where they prevent large debris fragments from entering the pump. It is a good idea to check the filter screen occasionally; remove and clean it with a stiff-bristled brush and warm, running water if it needs a cleaning. Replace a deteriorating filter screen. Also, it is important to note that a frequent washing of dishes with heavy food wastes —bones, coffee grounds, shell fragments, and so on—will probably damage the filter screen and could ruin the dishwasher. A dishwasher is not a garbage disposer and should not be used as one.

Finally, some dishwasher troubles are directly related to an improper loading of the unit. Always allow space between dishes and utensils for water and air to circulate and do not place a large plate or utensil where it can shield other dishes from the spraying water. Always separate dishes so they do not strike each other and cause noise and damage. Also, do not put extremely light or fragile items in the dishwasher since the force of the washing action from the spray may displace them and cause either noise or damage. As a last suggestion, make sure all cups, bowls, glasses, and other similar dishes are placed so they will not catch water and then retain it during the drying cycle. Any water in excess of about 4 ounces carried over into the drying cycle will cause improper drying.

Automatic Dishwasher Troubleshooting Chart

Trouble	Probable Cause	Remedy
Dishwasher does not work at all.	Timer is defective.	Check timer for continuity; replace if defective.
	Motor is defective.	Check motor; repair or replace if defective.
	Door latch switch is defective.	Check door adjustment and continuity through switch.
	Timer switch or relay not energized.	Try switch or relay for action; replace if defective.
	Open circuit in wiring.	Check wiring for continuity; make sure all terminals are okay. Also check for loose leads. If any are found, secure them.
Dishwasher does not make a complete cycle.	Timer operating erratically or improperly.	Check timer, replace if defective.
	Loose or broken lead.	Check and secure or repair all faulty leads.
	Motor overload is defective.	Check overload protector; replace if defective.
	Interlock switch is opening circuit.	Check interlock switch; must replace if defective.
	In models with thermostat, thermostat is defective or not making contact with tub.	Check thermostat for continuity; replace if defective.
Too little or no water in dishwasher.	Flow washer is defective.	Check and install new washer.
	Fill valve screen or fill valve is blocked.	Check filter screen and fill valve for foreign material.
	Inlet valve defective or inlet valve solenoid burned out or shorted.	Check for continuity and shorts; replace valve if defective.
	Pressure switch is defective.	Check; replace if defective.
	Timer is defective.	Check for continuity; replace if defective.
Water does not drain out.	Pump is blocked or inoperative.	Clear restriction; replace motor or pump if necessary.
	Drain valve is defective.	Check drain valve or solenoid; repair or replace as necessary.
	Timer is defective.	Check; replace if defective.
	Drain screen is clogged.	Clean or replace screen.

Automatic Dishwasher Troubleshooting Chart (Continued)

Trouble	Probable Cause	Remedy
Water does not drain out.	Broken impeller blade in pump.	Pump must be replaced.
	House plumbing is blocked or backing up.	Check and take necessary action.
Dishwasher leaks.	Loose plumbing connections.	Check all hose connections; tighten or repair.
	Pump is leaking.	Replace valve seal assembly, if necessary. If housing is cracked, replace pump.
	Fill valve overspray or oversplash.	Check fill valve for misalignment, burrs, or deposits left by hard water.
	Excessive foaming.	Improper detergent, or too much detergent in a soft water area.
	Worn or damaged door gasket.	Check installation; replace gasket if defective. Also check for loose door latch.
	Drain valve leaks or sticks open.	Clean all drain valve components.
Dishwasher is noisy.	Chattering noise.	Check fill valve diaphragm.
	Foreign object in pump or impeller blade.	Remove and check pump and impeller blade; if broken, bent, or unbalanced, repair or replace.
	Spray arm hitting.	Check clearances; make sure no dishes are in the way of the spray arm. Check hub and nut to be sure they are tight.
	Foreign matter in tub.	Check for broken dishes or similar objects.
	Loose parts.	Tighten as necessary.
	Low water level.	Water should submerge the heater.
Dishes do not dry properly.	Improper loading.	Dishes should not be overcrowded, preventing air circulation.
	Low water temperature.	Water in dishwasher should be 140 to 160 degrees F at last rinse. Adjust heater thermostat accordingly.
	Timer contacts not closing during heat cycle.	Check timer; replace if defective.
	Heater does not work.	Check heating element for continuity; replace if defective.
Dishes not clean.	Improper water level.	Check service manual for proper water level; normally, heater should be submerged.
	Spray arm not moving.	Check for binding. Check for clearance between arm and bracket, rail, and so on.

Automatic Dishwasher Troubleshooting Chart (Continued)

Trouble	Probable Cause	Remedy
Dishes not clean.	Impeller is defective.	Check; replace if inoperative.
	Low water temperature.	Water temperature during last rinse should be 140 to 160 degrees F.
	Loose or dirty filter screen.	Refit filter screen to eliminate gaps; wash out thoroughly.
	Pump is defective.	Check; replace pump if necessary.
	Drain line is plugged.	Clean out drain line.
	Improper loading of dishes.	Pots and pans should not be placed on top of other dishes. Cups, glasses, and other dishes that catch water should be upside down.
	Improper use of detergent.	Use detergent formulated for use in automatic dishwashers; use recommended amount.
	Dishes prepared improperly.	Dishes should at least be scraped before being put in dishwasher.
Dishwasher blows fuse or trips circuit breaker.	Short circuit.	Check components for continuity; start with line cord connector, a common problem area.
	Motor overloaded.	Check for object blocking or jamming pump blade.
	Other appliances on the same circuit may be causing overload.	Check; if necessary, put dishwasher on a separate circuit or put some of the other appliances on a different circuit.

Electric Ranges and Ovens

No other category of appliances displays the variety that electric ranges and ovens do in sizes, configurations, and features. But all electric ranges are alike in having two separate sets of heaters: the surface units and the oven unit or units. The more popular configurations are the freestanding range, the built-in oven, the slide-in oven, and the drop-in oven. We will also briefly look at two smaller cousins of the general oven family: the microwave oven and the convection oven.

The freestanding range as its name implies, stands by itself, independent of a wall or cabinets on either side. It can be a single-oven model with a broiler or a double-oven with an eye-level oven or a broiler, or even with a microwave oven. The controls may be in the front, on top, along the side, or on a backsplash. The built-in oven, with either a single or double cavity, must be fitted into a wall cabinet built for this purpose, and as an added feature it sometimes comes with a microwave component. This built-in oven also frequently features a built-in cooktop or surface unit range instead of separate heating elements. The slide-in oven is basically a freestanding range, but with the side panel left off and engineered to fit snugly against the countertop or even overlap it for a built-in look. The slide-in ranges are available in the same configurations as the freestanding styles. A variation of the slide-in, the drop-in does not rest on the floor. Flanges rest on the countertop on either side, and it is supported from there. Special cabinets are available to fit under it for a more built-in look, although it might extend all the way to the kick space.

Regardless of the physical construction of the various ranges and ovens, their operating principles are essentially the same. All traditional electric ranges use units (resistances) to produce heat for cooking. Depending upon unit design, this heat is applied through conduction, convection, or radia-

A modern oven featuring surface units for frying, melting butter, boiling, simmering, and sautéing, as well as oven units for baking, broiling, and roasting.

tion. For instance, all top or surface units are designed with a large amount of surface area for greater conduction. Oven bake units use the principle of convection for heat transfer. These units are designed to fit the oven cavity and occupy a minimum of space. Air circulation around the unit is most important. On the other hand, the basic function of a broiling unit is to emit large amounts of radiant energy or radiation uniformly over the surface of the broiler grill. This usually requires longer heating coils. Most ranges locate the broil unit in the top of the oven cavity.

The principle of operation of ranges and ovens is basically the same as that in all heat-producing appliances, large or small. Heating is achieved by connecting a resistance of the proper amount across an electric potential, therefore creating current flow. Variations in heat are generally obtained by connecting two or more elements in parallel or series and/or by varying the amount of voltage supplied to the heating units. Control of these units (resistances) to meet the needs of the various cooking requirements is by means of switches. The units and controls vary according to the requirements of the design and the control to be used.

Surface Units Modern surface unit—stove top—heating elements are comprised of a nichrome resistance wire which is insulated with magnesium oxide. This is placed inside of a small Inconel tube which forms a sheath. The wattage, or heat-producing capacity, of these surface units is directly related to the number and size of the wire coils which set up an ohmic resistance to the electric current. In most ranges there are two 6-inch surface units of approximately 1,250 to 1,500 watts each, and two larger 8-inch units of about 2,000 to 2,800 watts. A few range models feature a 4-inch surface unit. The heating elements are usually supported in the range top by a trivet that rests on a trim ring partially recessed into the range top surface. Drip bowls of porcelain on steel, aluminum, or stainless steel catch boilovers or spilled foods. The element may be of the hinged style or the plug-in style. The hinge mounting bracket on surface units usually includes provisions for standing the element up so that the drip bowl can be cleaned. Plug-in units on the other hand can be completely removed from the range to facilitate cleaning under the surface unit.

Surface heating units use three principal types of control switches to regulate temperature: fixed-position switches, infinite-heat switches, and automatic controls. Fixed-position switches were commonly used in older model surface units; they are now used much less often. Most of the ranges that use a fixed-position switch today use a seven-heat type which utilizes dual-resistance units of unequal wattage by taking advantage of the two voltages—240 and 120 volts—available in a three wire 240-volt system. Between the two resistance units, singly or in combination and utilizing the two available voltages, seven cooking temperatures are available. Usually these temperature positions will be high, medium high, medium, medium low, low, very low, and simmer. The disadvantage of the fixed-switch system is that you are locked into using seven set temperatures with no fine variances possible.

The surface units with an infinite-heat switch are heated to their full-rated wattage whenever the switch is making contact. On the high setting the switch keeps the surface unit on all of the time. At a lower heat setting the unit is cycled on and off by the switch. In other words, at the high setting the surface unit is energized continuously; at the warm setting the unit is energized only about 6 percent of the total time it is on. The infinite-heat switch gets its name because it can be set at any of an infinite number of settings between the high and warm positions. The ON-OFF cycling is accomplished by a bimetal strip in the switch.

An automatic control constantly provides the temperature selected on the control dial. The sensing element, mounted in the center of the surface unit, measures the temperature of the utensil on the element and regulates the amount of heat accordingly. The control is essentially a hydraulic thermostat. A bimetal heat anticipator inside the control opens and closes the cycling switch by means of heat produced by the surface unit current that it carries. This portion of the control functions only when the temperature of the hydraulic system has reduced the contact pressure, which occurs very close to the preset temperature of the control dial.

The most common and increasingly popular variations of the regular surface unit are the glass cooktops of a ceramic-type material which are now being used as surface units for ranges of all types and built-in counter units. In most of these cooktops, heat is applied to the ceramic cooking panel by any one of or all four heating elements mounted under the surface and properly positioned by element retainer bars. The great advantage of the ceramic-glass cooktops is that they are much easier to clean than traditional surface units.

Oven Units There are three basic types of oven units in use today: the conventional oven, the continuous-cleaning oven, and the self-cleaning oven. Actually, there are two other distinct types of ovens, the microwave oven and the convection oven, but they generally come in portable models and are discussed separately for that reason. Modern ovens incorporate automatic control devices, such as automatic food timers and temperature gauges, to a much greater extent than surface units, which still usually require a lot of close attention from the cook when they are being used.

Most conventional or non-self-cleaning ovens employ tubular heating elements of the same design and materials as the surface units. Generally two tubular heating elements are used: a broil element located at the top and a bake element located at the bottom. Both elements operate at 240 volts and require anywhere from 2,000 to 4,000 watts, depending on the model. Usually, you choose your function by flipping a switch on the range or oven.

The temperature control circuit in a conventional oven consists of an oven control thermostat in series with a selector switch. Once the user sets the selector switch to the desired cooking operation and turns the unit on, the thermostat senses the temperature in the oven, and then by opening and closing the heating element circuit, maintains a relatively constant oven temperature for the cooking operation selected.

Of course, what makes the conventional oven different from the continuous-cleaning oven and the self-cleaning oven is that you have to clean its interior by hand. Continuous-cleaning ovens require no special insulation, vents, controls, or thermostats and are cleaned at normal oven temperatures by the process of organic oxidation during normal usage. What makes them clean continuously is the type of porcelain used on the inside of the oven. Porcelain used on both the conventional ovens and the pyrolytic self-cleaning ovens is very smooth; the surface might be compared with the texture found on a glass pane. The porcelain finish used in the continuous-cleaning oven has a texture that is more comparable to that of a brick. It is dark in color and contains many white stipples. When a fat spatter strikes regular porcelain, it tends to form a drop just as water or liquid striking a piece of waxed paper would form a drop. The drop of fat or food dries, causing the oven to soil. On the porcelain in the continuous-cleaning oven, the spatter is dispersed in much the way that a drop of moisture is in hitting a paper towel. With the spatter broken down into small particles in this way, it is decomposed at normal oven temperatures. The continuous-cleaning surface works well on fat spatters —the soil that generally occurs during roasting, broiling, and oven frying. Spatters created by spillovers from pies and casseroles are not decomposed on the continuous-cleaning finish. This type of mess generally collects at the bottom of the oven, and many manufacturers recommend the use of a heavy foil liner on the bottom of continuous-cleaning ovens to collect the refuse.

The self-cleaning oven makes use of a process called pyrolysis. The pyrolytic method of oven cleaning involves the chemical decomposition of cooking soils which consist primarily of compounds of hydrogen and carbon present in cooking oils, greases, sugar, starch, and so on. This is promoted by the application of extremely high heat. The process begins to occur at temperatures above 700 degrees F. The temperature required for a satisfactory cleaning job at a reasonable speed is in the area of 850 to 1,050 degrees F; the top limit depends on the model.

Normally, the cleaning process requires that the high-heat be held for 2 to 4 hours. Actually, the main difference in controls from conventional oven controls is a design to withstand these temperatures and control them accurately; that is, range models with self-cleaning ovens can, in most instances, be considered standard ranges from the service viewpoint. They have the additional feature of self-cleaning, but basically use the same components to bake and broil. During the cleaning operation the circuitry is changed and a few extra components are added to control the cleaning. The three basic types of temperature controls used in conjunction with the self-cleaning operation are a bimetal-type thermostat, a responder-type control circuit, and a solid-state control. Other components that are normally part of the oven self-clean system are a latch assembly which makes sure that the oven door cannot be opened during the high-heat cleaning periods, a fan assembly to channel hot air up both sides of the range and release it from surface unit openings, and a smoke eliminator to handle smoke removal during the cleaning process.

Microwave and Convection Ovens Microwave and convection ovens are usually either portable or in any case rather small countertop ovens that can perform many of the same tasks as a traditional oven, although they both go about these jobs somewhat differently. When food is prepared in a traditional oven, the heat surrounds the food and gradually penetrates to the center to accomplish its work. When food is prepared on a range surface unit, heat is applied from an electric coil to the food container, such as a frying pan. Heat then penetrates into the food being cooked by conduction. In the microwave or electronic oven, however, foods are heated by the absorption of very short electromagnetic waves which cook foods very quickly; an average cooking time is about 10 minutes per pound for most meat cuts. The wattage output for these mini-ovens ranges from 300 to 700 watts; however, most microwave cookbooks are written for 600- to 700-watt models, which will, of course, cook food about twice as fast as a 300-watt model. Microwave ovens frequently offer a variety of extra features including a digital timer, an automatic cycling defroster, a variable power dial for changing cooking speed, automatic food temperature controls to stop the oven automatically when the internal temperature of a cooked food reaches the temperature you have set, and an optional browning coil.

Microwave oven malfunctions other than the usual circuitry problems usually involve a defective latch or interlock switch, a faulty timer, or a defective thermal overload protector. A magnetron tube is the heart of a microwave oven, as it converts DC power to radio frequency power or microwave energy. The filament in the magnetron tube can develop an open circuit, or the internal plate cathode can short out. In either case, the tube, which must be handled very carefully, will have to be replaced. Also, a microwave oven should be kept clean to prevent odors from building up. To clean, use a mild detergent, water, and a soft cloth. Odors inside the oven can be eliminated by boiling a solution of one cup of water and several tablespoons of lemon juice in the oven for five to seven minutes.

Convection ovens, on the other hand, use electrical energy as do traditional ovens, but they make more efficient use of the heat produced. The air inside of a traditional oven is almost static during its operation, and cooking depends on the gradual conduction of heat from the outside to the middle of the food. Convection ovens, however, make use of a stream of air, produced by a high-speed fan, that swirls continuously over a standard heating element while the unit is on. This results in a uniform temperature throughout the oven, which not only speeds up the cooking process, but saves up to 50 percent of the energy used by traditional ovens. Therefore, foods can be cooked at lower temperatures for shorter periods of time in convection ovens. For example, baking temperatures may be lowered by as much as 75 degrees F and roasting times can be cut by as much as a third. Because convection ovens give off less heat than traditional ovens, the kitchen will be cooler when they are used. Their disadvantage is that they, like the microwave ovens, are too small for a lot of big cooking jobs. They are especially easy to clean because of their compact interiors. Most problems associated with their operation are similar to those found in regular ovens, except for any trouble you might have with the fan or fan motor.

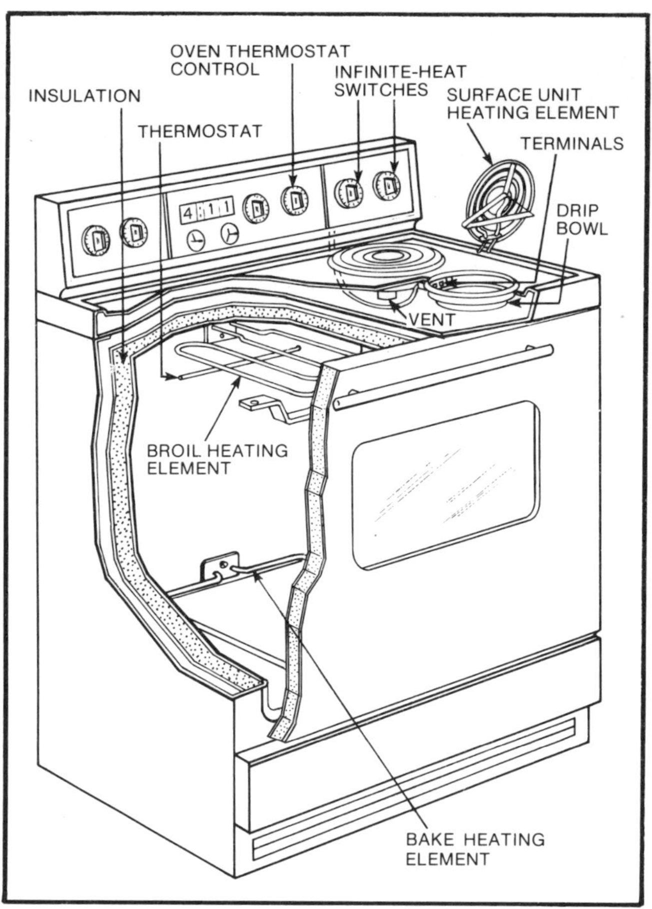

Cutaway view of a typical oven/range unit.

Repairs and Servicing Problems involving an oven or range can be caused by a wide range of malfunctions, but one which commonly occurs is a heating element or elements failing to get hot. If a surface heating element does not heat up, the first step in checking it is to be sure that power is being supplied to the range. This can be determined quickly by checking to see if any other heating elements on the surface unit or in the oven heat. After you are sure that the range is receiving power, the element can be unplugged from the unit. Set the surface control for that unit on the HIGH setting. With the VOM set on the 250- or 300-volt scale, check for voltage at the surface unit terminal block. If there is voltage at the terminal block, then check the surface element for continuity. Set the selector for an ohm or resistance reading and zero the dial. The resistance should be very low, generally 50 ohms.

Some elements have two or three heat patterns. Be sure to check the resistances of each pattern for continuity by applying the VOM probes to every other terminal. If the element shows no resistance, it should be replaced. Also, if for any reason the element shows extremely high resistance, it should be replaced because it may have a short.

To check the heating elements in an oven, first observe the elements when they are hot for even redness throughout their heated length. This check must be made with the oven control set on BAKE when checking the bake unit and on BROIL when checking the broil unit. Most oven heating elements are completely front-servicing; that is, both the broil and bake elements may be removed from the inside of the oven by removing a screw at each side of the element and pulling the element out of the oven cavity. Be careful not to pull too hard when removing the oven element because it is possible to accidentally disconnect the spade terminals if care is not taken. Bake and broil elements have a very low resistance and should be checked at the terminals in the same manner as surface units.

A common complaint with ovens is that they do not heat properly. You will need a good temperature tester or thermocouple to accurately check the oven's temperature. To check the heating ability of the oven, place the thermocouple lead in the center of the oven. Turn on the oven and set the thermostat at 400 degrees F. Allow the oven to cycle four times or wait 20 minutes before taking a reading to allow the temperature in the oven to stabilize. The average temperature of the high-low readings must fall between 390 to 430 degrees F. If the temperature does not fall within this range, you know that either a heater wire or the oven thermostat is defective. These components can be removed and checked individually for a malfunction.

Most no-heat complaints in an oven that can be traced to the temperature control circuit are generally caused by problems with either the selector switch or the thermostat. If the proper contacts are not closed in either of these components, current cannot flow to the elements and relays. Because of the variety of selector switches and thermostats, you will have to check a wiring diagram to determine which terminals to check. But before replacing an oven unit control switch, closely inspect the lead and quick-connect terminals for corrosion or thermal breakdown. Each terminal should be clean and fit tightly on the switch. Any terminal that does not meet these standards should be replaced. Should the wire be oxidized or show signs of overtemperature, it should be cut back before replacing the terminal. The entire lead assembly should be replaced if severe oxidation is evident. Because of the high electric currents encountered in the operation of an oven, it is important that all wire splices and quick connects are properly soldered.

Electric Range and Oven Troubleshooting Chart

Trouble	Probable Cause	Remedy
All surface elements do not heat.	Fuse blown or circuit breaker tripped.	Find what caused problem; short circuit could be in range. Correct any problems and change fuse or reset circuit breaker.
	Loose, broken, or burned leads or connections.	Check visually; then make continuity tests. Repair as necessary.
Single surface element does not heat.	Heating element is burned out.	Check for continuity; if open, replace element.
	Terminal contacts are bad.	Replace terminal block.
	Wires loose or burned off.	Tighten or replace wires.
	Control switch or infinite-heat switch is defective.	Replace switch.
All surface units do not heat (ceramic cooktop only).	Safety circuit open.	Ceramic top is cracked. Replace top and fuse links as needed.

Electric Range and Oven Troubleshooting Chart (Continued)

Trouble	Probable Cause	Remedy
All surface units do not heat (ceramic cooktop only).	No power to cooktop.	Check fuses or circuit breaker; then check continuity along wiring to element terminal.
Oven does not heat.	Fuse blown or circuit breaker tripped.	Find what caused problem; short circuit could be in oven. Correct any problems and change fuse or reset circuit breaker.
	Timer set wrong.	Reset timer correctly.
	Timer is defective.	Check timer; replace if defective.
	Thermostat is defective.	Check thermostat and replace if defective.
	Oven heating elements are defective.	Check elements for continuity; replace if defective.
	Selector switch is defective.	Check switch; replace as needed.
Oven does not get hot enough or gets too hot.	Thermostat calibration is incorrect.	Recalibrate thermostat if possible.
	Thermostat is defective.	Check; replace if defective.
	Improper door fit.	Check and take corrective action.
	Altitude above sea level.	Temperature calibration is about 4 degrees F low for each 1,000 feet of elevation from sea level.
Timer does not control oven.	Incorrect timer setting.	Use manufacturer's guide book for timer settings.
	Timer is defective.	Replace timer.
	Timer motor is defective.	Check timer motor; if defective, replace motor or timer if that is more feasible.
The oven does not bake.	Bake heating element is defective.	Check bake element for continuity; replace if defective.
	Thermostat is defective.	Check thermostat; replace if defective.
	Selector switch or wiring defective.	Check selector switch and wiring; replace defective parts.
The oven does not broil.	Broil heating element is defective.	Check broil element for continuity; replace if defective.
	Thermostat is defective.	Check thermostat; replace if defective.
	Selector switch or wiring defective.	Check selector switch and wiring; replace defective parts.
No heat for the cleaning operation (self-cleaning ovens only).	Control setting incorrect.	Use manufacturer's instructions for cleaning procedure.
	Selector switch is defective.	Replace switch.
	Oven door is not locked.	Correct situation.
	Thermostat is defective.	Check thermostat; replace if defective.
	Oven relay defective.	Check and replace relay.

Electric Range and Oven Troubleshooting Chart (Continued)

Trouble	Probable Cause	Remedy
No heat for the cleaning operation (self-cleaning ovens only).	Broil heating element is defective.	Check for continuity; replace if defective.
	Timer switch is open.	Check; replace if necessary.
	High-limit switch is defective.	Replace switch.
	Open interlock switch (if used).	Replace switch.
	Open pulse switch (if used).	Replace switch.
	Open bias circuit (if used).	Repair or replace bias control.
Incomplete cleaning (self-cleaning ovens only).	Control setting incorrect.	Use manufacturer's instructions for cleaning procedure.
	Timer set for too short a time period.	Most models require at least 2 hours for cleaning.
	Oven is excessively dirty.	Remove some of the excess build-up manually before starting cleaning cycle.
	Heating element or elements are not working properly.	Check each for operation in bake and broil position. Replace defective elements.
	Oven temperature control calibration for cleaning is too low.	Check oven temperature; adjust or replace oven temperature control as necessary.
	Bias circuit set too high (if used).	Recalibrate.

Garbage Disposers

There are two basic advantages in having a garbage disposer: sanitation and convenience. With a garbage disposer you can get rid of your food wastes immediately with no mess and no scraping. In fact, some cities have ordinances requiring disposers in all new units being constructed. There are presently two basic types of garbage disposers: the continuous-feed type and the batch-feed type. Regardless, all disposers today are designed to fit any sink that has a standard drain opening of 3½ to 4 inches in diameter.

The more popular model is the continuous-feed type of disposer, which does just what its name implies—it accepts garbage continuously. You just turn on cold water, flip a conveniently located single-pole disposer switch, and feed waste into the sink opening. A flexible rubber covering or backsplash baffle covers the disposer opening, and the food waste is simply pushed through the baffle. Actually, this baffle has four distinct purposes: It isolates the disposer from the sink fitting; it maintains a curtain of water over the opening to baffle the noise of grinding; it prevents backsplash; and it helps to prevent foreign articles from falling into the disposer.

Garbage disposers can be either continuous-feed or batch-feed models.

The batch-feed or lock-cover type of disposer permits you to dispose of a batch of garbage at a time. Garbage is first fed into the disposer opening, which has no baffle. Then a metal cover is inserted in place and twisted. This activates the disposer.

When the batch of garbage has been disposed, the cover is turned in the opposite direction to shut the appliance off and to lift the cover out. Then another batch of garbage is inserted until all the garbage is gone.

All modern food waste disposers consist of four main components: an electric motor; two grinding elements, including the rotating shredder or flywheel and impeller plus the stationary grind ring; grinding chamber; and sink-fitting mounting arrangement.

Motor sizes normally range from ⅓ to ½ horsepower. The larger the horsepower, the more torque is created, reducing the incidence of jamming. Motors are generally either capacitor-start, which brings the rotating shredder up to full speed almost instantly, or split-phase, which takes longer to come up to speed. The split-phase motors also draw excessive amperage on house current when turned on. Most split-phase motors have a centrifugal starting switch to turn on the rotor, while the capacitor-start type usually employs a relay.

Generally, a primary seal is installed to prevent water and waste from entering the bearing and motor areas. A few manufacturers also use some sort of plastic protective shield against the corrosive action of detergents and electrolysis. This detergent shield is generally made of polypropylene and is resistant to electrolysis and corrosion. In some other makes, the grind chamber itself is given an epoxy coating.

A look at most disposer motor housings reveals a rotor shaft protruding with the rotating shredder or flywheel attached to it with a positive nonslip drive. A solid grinding or cutting element can also be attached and locked to the shaft, providing direct drive. A stationary grind ring is anchored into position by press-fitting. It has a predetermined internal clearance, allowing the rotating shredder or flywheel to revolve within its circumference. The internal clearance is equal to the thickness of a cigarette paper, providing a steady, consistent grind size. While reverse switches operate on different principles in different makes, all will reverse the disposer every time the wall switch is turned on. The feature doubles the life of the cutting elements.

Regardless of the type of motor utilized, the basic principle of operation for all garbage disposers is centrifugal force. As we already know, food waste is dropped through the opening into the grind area. With the disposer operating at approximately 1,700 rotations per minute, it throws food by centrifugal force against the walls of the grind chamber where the stationary grind ring is located. The lugs or impellers on the rotor shredder or flywheel channel the food waste into the grind ring. In essence, the waste is grated, almost as an onion is on an onion grater, until the particles are reduced in size. When they become small enough, they slip between the teeth of the grind ring and through the holes of the rotating shredder. While this principle of operation is basically the same for all disposer makes, some models have movable cutting impellers or lugs that lie flat, then rise into position when the disposer is turned on. The manufacturers claim various advantages for both types, but essentially they do the same job. When the particles are small enough to fall beneath the rotating shredder or flywheel, the water carries them out through the drain line.

One of the most important operating rules with any garbage disposer is that it must be used with a full flow of cold water only, never hot water. Cold water congeals grease into globules that flow easily through drain lines, while hot water makes grease more fluid and prone to coat the drain line. The cold water must be turned on before the disposer is made operative and left on for a full 30 seconds after the disposer is turned off. Also, disposers should not be used with a grease trap in the drain line. If present, the trap must be either removed or bypassed. If the grease trap has a removable baffle, its removal will solve the problem. Some disposers are self-cleaning, others are not, depending upon design and material. But it is a good idea to flush out all units occasionally, thereby ensuring their being odor-free. Flushing can be done by filling the sink with water, turning the disposer on, and removing the stopper. The action of the swirling water will remove residue food wastes.

Most food waste can be disposed of in a garbage disposer, as long as the waste fits all the way in. This would eliminate, for example, large and bulky bones. It is generally recommended, however, that stringy waste such as celery be fed slowly and with other forms of food waste at the same time. Some garbage disposers have great difficulty with waste that splinters and is hard to grind—small chicken bones, for example. If you find that your disposer does not do a good job with certain objects, avoid putting them in at all. Soft-surfaced paper, such as napkins and cleaning tissue, can also usually be eliminated in a disposer. However, the primary function of the disposer is to dispose of food wastes. It should not be expected to handle material such as bottle caps, metallic foil, rubber articles, broken glassware, tin cans, sponges, metal cleaning pads, oystershells, clamshells, string, towels, or dishcloths. If such material is inadvertently dropped into the shredding compartment, you should immediately lift out the sink stopper and splash guard or turn the disposer off and remove the material.

Any disposer, of course, will jam if a metallic object is accidentally dropped into it. Jamming of

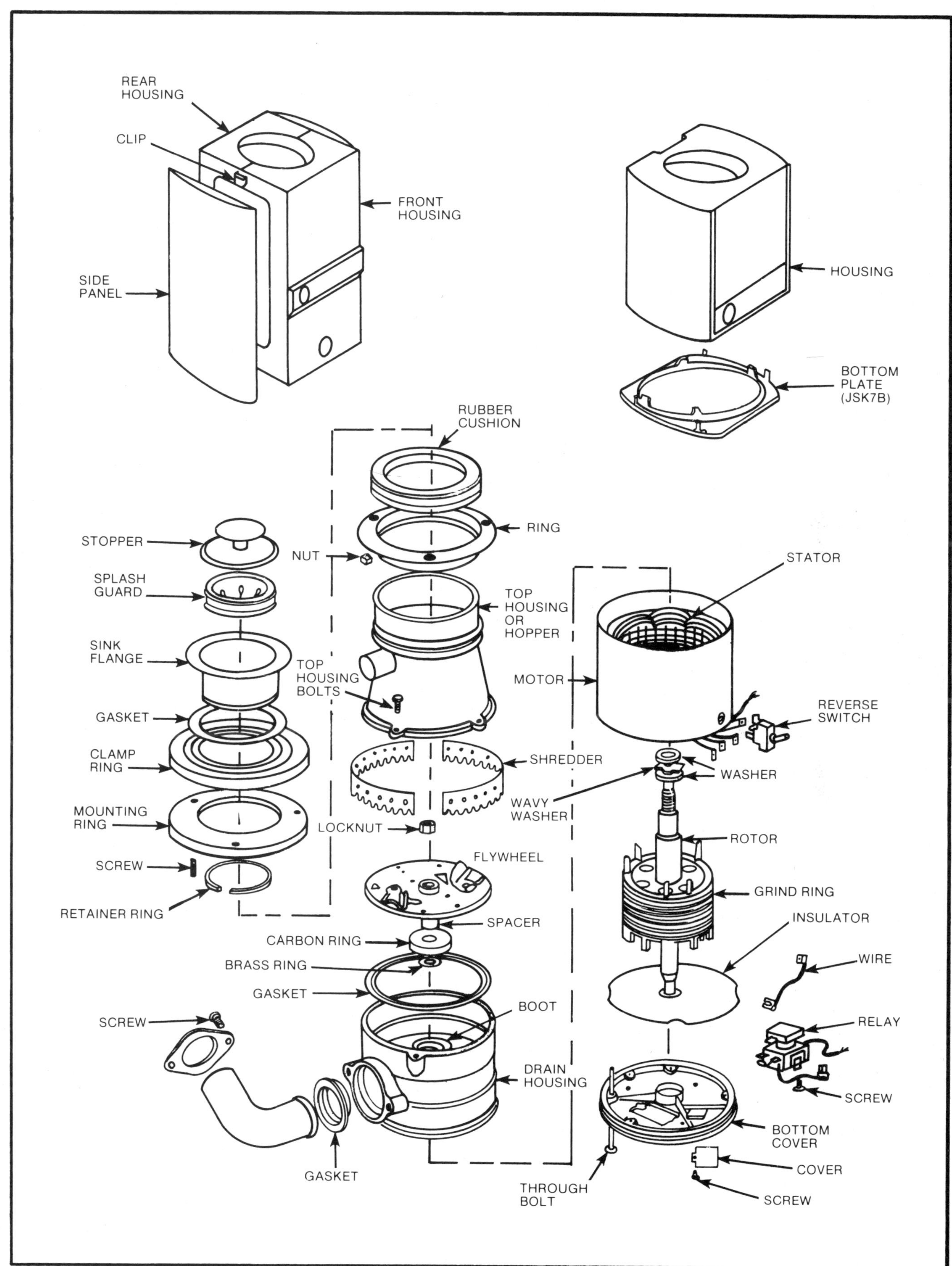

Exploded view of a typical home-model garbage disposer.

Garbage Disposer Troubleshooting Chart

Trouble	Probable Cause	Remedy
Disposer does not work and motor does not start.	Loose or broken wires at either ON or OFF switch, service connections at terminal plate, wire connections at overload protection switch, capacitor or start-reverse switch.	Check for continuity; replace defective wires. Resolder or reconnect broken connections.
	Defective start-reverse, start or relay switch. Contacts could be burned or welded together.	Replace switch or relay. Check contact points on switch or relay to be sure points make contact.
	Capacitor open, shorted, or weak.	Check capacitor for continuity; replace if defective.
	Motor is defective.	Repair or replace motor.
	Jammed condition has caused overload protector to cut off current.	Check for metal or other foreign object between the cutting elements. Use the proper jam release tool, if available. On some disposers it is necessary to reset the overload switch before the disposer will work.
	Housing broken on overload protector switch, or overload protector switch has cut out or burned out.	Push reset button. If disposer does not work, check wiring for continuity; then check overload protector switch for continuity. Replace defective parts.
Motor runs for a short interval and then stops.	Relay is defective.	Replace relay.
	Short in stator winding.	Replace stator.
Overload protector switch kicks out continuously.	Jammed condition has caused overload protector to cut off current.	Check for a metal or other foreign object between cutting elements. Use proper jam release tool, if available. With some disposer models it is necessary to reset overload switch.
	Capacitor open or shorted.	Check capacitor for continuity; replace if defective.
	Shorted or grounded winding on stator.	Replace stator.
	Overloaded circuit.	If possible, connect disposer to separate circuit.
	Broken or disconnected lead wires.	Check for continuity; repair as needed.
	Overload protector switch is defective—too sensitive.	In a locked rotor position, the overload protector switch should trip in an average of 10 seconds.
	Motor switch contacts burned or welded together.	Replace switch.
Excessive operating time to dispose of the waste load.	Undisposable material in grinding chamber.	Stop operation of disposer and remove material.
	Worn or broken tab on shredder ring.	Replace shredder ring or stationary grinding ring as needed.

Garbage Disposer Troubleshooting Chart (Continued)

Trouble	Probable Cause	Remedy
Excessive operating time to dispose of the waste load.	Insufficient flow of cold water.	Minimum cold water flow should be 1½ gallons per minute.
	Stuck or badly worn impellers.	Free impellers; replace rotor shredder or flywheel assembly if necessary.
	Dull or broken edges on the shredding knives.	It is impractical to sharpen the shredder knives; therefore, replace them.
	The waste consists of a great amount of hard, fibrous material or large bones.	Waste disposal can sometimes be speeded up by increasing the cold water flow or by cutting up material beforehand.
Disposer excessively noisy or causes great vibration.	Backup screws, mounting screws and nuts, or other hardware is loose.	Tighten nuts and screws.
	Undisposable material in disposer.	Turn off disposer and remove material.
	Broken rotor shredder or flywheel.	Replace damaged part.
	Motor bearings are damaged or make squealing sound because they need to be lubricated.	Disassemble and lubricate; replace if damaged.
	Rotor shredder or flywheel striking stationary grind ring because of broken thrust washer.	Replace washer.
Water leaks around sink or strainer flange.	Mounting is loose.	Tighten all appropriate screws.
	Gasket is defective.	Replace gasket.
Water drains slowly or not at all.	Drain line or opening is blocked or partially blocked.	Close sink and fill it with water; turn on disposer and lift sink closure. Water will be forced into drain, cleaning it out. If you have an adjacent sink, make sure its drain is closed during this operation.
	Insufficient flow of water during disposer operation.	Open water faucet for a heavier flow of water.
	Worn cutting elements not grinding waste fine enough.	Remove sink trap and place empty pail under outlet of disposer. Grind available waste and check size of ground particles in bottom of pail. If waste is passed in very large chunks, or if potato peelings, lettuce leaves, and so on have not been ground, replace cutting elements.
Foul odor from disposer.	Improper operation of disposer allowing food particles to remain in unit and decompose.	Allowing motor and cold water to run at least 30 seconds after disposing of waste will thoroughly clean out disposer.

Garbage Disposer Troubleshooting Chart (Continued)

Trouble	Probable Cause	Remedy
Disposer blows fuses or repeatedly trips circuit breaker.	Short circuit.	Make continuity tests on electrical circuits, especially stator winding, until short circuit is found.

the flywheel or rotating shredder may also be caused by shutting off the cold water or the wall switch before the waste has been entirely disposed of. The motor may come to a stop with slivers of bones or other hard material wedged between the flywheel and the stationary shredding knives. When the motor is later restarted, the flywheel or rotating shredder may not immediately develop sufficient impetus to cut through the obstruction, with the result that the motor stalls.

Overfilling may cause jamming, too. It is better to dispose of waste in small quantities than to accumulate a large amount. The flywheel or rotating shredder may also jam if the flow of water to the unit is obstructed. For example, if the sink is loaded with pots, pans, and dishes so that water runs into these articles instead of into the drain, the flywheel may jam and stall the motor before sufficient water finds its way to the drain. One final note should be made about servicing garbage disposers. Be extremely careful when attempting to unjam a disposer. Be sure you open the circuit to the disposer at the service box before you start tinkering with your disposer unit.